THE WILL TO TECHNOLOGY AND THE CULTURE OF NIHILISM: HEIDEGGER, NIETZSCHE, AND MARX

D0913618

In *The Will to Technology and the Culture of Nihilism*, Arthur Kroker explores the future of the twenty-first century in the language of technological destiny. Presenting Martin Heidegger, Karl Marx, and Friedrich Nietzsche as prophets of technological nihilism, Kroker argues that every aspect of contemporary culture, society, and politics is coded by the dynamic unfolding of the 'will to technology.'

Moving between cultural history, our digital present, and the biotic future, Kroker theorizes on the relationship between human bodies and posthuman technology, and more specifically, wonders if the body of work offered by thinkers like Heidegger, Marx, and Nietzsche is a part of our past or a harbinger of our technological future. Heidegger, Marx, and Nietzsche intensify our understanding of the contemporary cultural climate. Heidegger's vision posits an increasingly technical society before which we have become 'objectless objects' – driftworks in a 'culture of boredom.' In Marx, the disciplining of capital itself by the will to technology is a code of globalization, first announced as streamed capitalism. Nietzsche mediates between them, envisioning in the gathering shadows of technological society the emergent signs of a culture of nihilism. Like Marx, he insists on thinking of the question of technology in terms of its material signs.

In *The Will to Technology and the Culture of Nihilism*, Kroker consistently enacts an invigorating and innovative vision, bringing together critical theory, art, and politics to reveal the philosophic apparatus of technoculture.

(Digital Futures)

ARTHUR KROKER is Canada Research Chair in Technology, Culture, and Theory at the University of Victoria.

The Will to Technology and the Culture of Nihilism

Heidegger, Nietzsche, and Marx

ARTHUR KROKER

UNIVERSITY OF TORONTO PRESS
Toronto Buffalo London

© University of Toronto Press Incorporated 2004
Toronto Buffalo London
Printed in Canada

ISBN 0-8020-8786-8 (cloth)
ISBN 0-8020-8573-3 (paper)

Printed on acid-free paper

National Library of Canada Cataloguing in Publication

Kroker, Arthur, 1945–
The will to technology and the cuture of nihilism : Heidegger,
Nietzsche, and Marx / Arthur Kroker.

(Digital futures)
Includes bibliographical references and index.
ISBN 0-8020-8786-8 (bound)   ISBN 0-8020-8573-3 (pbk.)

1. Technology and civilization.   2. Nihilism (Philosophy).
3. Heidegger, Martin, 1889–1976.   4. Nietzsche, Friedrich Wilhelm,
1844–1900.   5. Marx, Karl, 1818–1883.   I. Title.

CB478.K762 2004   303.48'3   C2003-904452-1

University of Toronto Press acknowledges the financial assistance to its
publishing program of the Canada Council for the Arts and the Ontario
Arts Council.

This book has been published with the help of a grant from the Canadian
Federation for the Humanities and Social Sciences, through the Aid to
Scholarly Publications Programme, using funds provided by the Social
Sciences and Humanities Research Council of Canada.

University of Toronto Press acknowledges the financial support for its
publishing activities of the Government of Canada through the Book
Publishing Industry Development Program (BPIDP).

Contents

Acknowledgments

As always, the writing of this book took place in the context of an ongoing intellectual conversation with Marilouise Kroker, who, in addition to co-authoring several of the stories in 'The Digital Eye,' has an unequalled understanding of the cultural consequences implicit in the unfolding future of the posthuman.

The completion of this book has taken place in the stimulating intellectual environment of the University of Victoria, which is very much a gateway to understanding the future of technology, culture, and theory in the twenty-first century.

My exploration of the will to technology within and beyond the writings of Heidegger, Marx, and Nietzsche owes much to Dr Stephen Pfohl, who, as Chair of the Department of Sociology at Boston College, established a creative learning and teaching context for revisiting the 'question of technology' and to the intellectual support provided by Dr Timothy Murray, Acting Director of the Virtual Realities seminar at Cornell University's Society for the Humanities.

I am deeply grateful to the Social Sciences and Humanities Research Council of Canada for research grant assistance which was vital to the completion of this book. I am also appreciative to the Canada Research Chairs program for my appointment as a CRC in Technology, Culture, and Theory at the University of Victoria, a form of truly innovative support that has made an interdisciplinary project of this order possible.

Finally, I would like to thank Chris Bucci, formerly of the University of Toronto Press (UTP), and Siobhan McMenemy, Frances Mundy, and Judy Williams of UTP for their care and diligence with the editing and production of the book.

The Culture of Nihilism

1 The Will to Technology

Third-Wave Eugenics

The inspiration for this book comes from a nomadic series of seminars on the digital future and the ethics of biotechnology, which I taught over the past several years at Cornell University, Boston College, and Concordia University. In each case, the focus was directly technological. The various readings served as probes of technoculture, sometimes technology as its rides the surface of the body in the form of digital media meant to amplify and extend the human sensorium, and, more urgently, technology as it invades the surface of the body, colonizing, coding, and manipulating the human genetic code. In that strange, ancient ritual by which the collective act of teaching and mutual learning sometimes mirrors the animating historical currents of its times, this series of seminars quietly, but no less decisively, effected a radical change in my intellectual autobiography. I began with an eagerness to explore the breaking wave of new digital media for their aesthetic, social, and political implications, but as the seminars proceeded I found my thinking on the 'question of technology' undermined and my predisposition to theorize new media at the speed of (aesthetic) light abruptly flipping into its opposite – a desperate need to think the future somehow outside, beyond, and anterior to the question of technology. Perhaps it was listening to students who spoke poetically and passionately about their deep feelings of 'mourning' for that which had been swept away by the bright sunshine of biogenetics. Maybe it was meditating upon the insistent demand for thinking ethically in the face of digital technologies that wired individual consciousness into the vacuity of a 'global brain,' and biotechnologies intent on harvesting human flesh by a historically momentous fusion of artificial intelligence and genetic sequencing.

Undoubtedly, this interior demand to intensify my thought was motivated by those strange accidents that always turn the mind most strongly: that winter of meditating upon the neglected texts of Heidegger, particularly his 1929–30 reflection, *The Fundamental Concepts of Metaphysics*, in the Graduate Library at Cornell University, teaching the digital future by day and listening by night to Heidegger's voice saying in the midst of the slaughterhouse of the already posthuman century that we cannot think the question of technology technologically. Or perhaps it was the realization during a seminar on ethics and biotechnology at Boston College, in the midst of a technologically dynamic city that likes to describe itself as 'Gene City,' that the question of the life sciences – the delirious spectacle of clonal propagation, genetic sequencing, organ farms, recombinant DNA, and artificial life-forms created out of the vivisectioning of plants, animals, and humans – cannot be thought outside of the question of nihilism. Listening to the euphoric prophecies of a new eugenics by the masters of the biotech universe, I was reminded of that other prophecy made by Nietzsche in *The Will to Power*. For Nietzsche, the twenty-first century was only the mid-point in a two-hundred-year cycle of nihilism, a storm of nihilism that would be equal parts passive ressentiment and a suicidal will to nothingness. Or maybe it was that report on the Web which told the story of the occupation of Monsanto's experimental 'seed farm' in Brazil by members of the Landless Peasants' Movement during the days of the Global Social Forum which made me turn once again to the question of virtual capitalism, going back to the texts of *Capital* for insight on the ascent of the 'digital commodity-form' from the now superseded era of industrial capitalism.

Who is to say what constitutes the genesis of an idea or what represents its most faithful moment of reconstruction after the fact? But one thing is certain. It happened. In theorizing the digital future and in teaching the question of ethics and biotechnology, I could not disentangle the question of nihilism from the instrumentalist explanation that the present drive towards the biogenetic future simply represents a necessary, technical extension of the digital and that the ethical questions raised by recombinant genetics could be thought satisfactorily only in terms of facilitating human health, 'improving' the human species, or prolonging the life of individuals. For myself, at some deeply informing emotional level my thought had already recognized, in the hysterical symptoms of biogenetics, a 'difference' that could not be recombined or willed away. The dynamic language of futurism expressed first by the utopian spin of new digital media and later by the life science industries and the newly emergent genetic class had about it the smell of something very ancient in the western tradition. In these visions recombinant of transgenic bodies, phosphorescent skin, jellyfish monkeys, firefly organs, mutant fish, sterile hybrid seeds, cross-species organs, there was the awakening again of the siren-call of a society intent on its own suicide, celebrating its coming disappearance in the language of the genetic modification of the species. Our biogenetic future might have begun with the naïve naturalism of Darwin's *The Origin of Species*, but the ending of the species may also be marked by writing which in the face of the recombinant future clings to a stubborn intellectual fact: it is only by thinking technology outside the horizon of the technological imperative, only by thinking biogenetics beyond the perspective of genetic determinism, that we can approach an understanding of *nihilism* as the essence of technological destiny.

Suiciding Itself to *Virtual* Life

It would be inaccurate to say that ours is merely a civilization of technological hubris. Nietzsche was more insightful. For him, we are a 'gamble,' a 'going-across,' a 'glance,' a 'gathering storm.' So was perhaps Jean-François Lyotard, for whom we are an 'incommensurability,' an impossibility that cannot be unrealized because we are perhaps never capable of full self-consciousness. Whatever the case, it can never and probably will never be said of us that we have not worn the membrane of technology as our deepest primal, that the horizon of technology is not the gamble upon which we stake the meaning of life itself. While it is a matter of strict epistemological warfare between social constructionists and hypermoderns to make much of the cultural issues attendant upon the meaning of the 'post' – post-society, post-culture, post-gender – I do not think that we have yet grasped deeply into the interstices of our thought just how graphically, how bleakly, we truly have become a culture of the post.

I do not mean this lightly. When the United States used nuclear weapons on Japan it precipitated a threshold event of the greatest cultural magnitude. History literally ended. If by history we mean the traditional cultural understanding of history as an indeterminate sense of unfolding time, an open future never fully under human control, then that sense of history decisively ended in the bio-flash of Nagasaki and Hiroshima. And again, Lyotard was right, but perhaps not in the sense he would wish. Post-history has been 'driftworks,' an indeterminate and increasingly violent series of technological experiments on the horizon of existence itself: the acceleration of space under the sign of digital culture until space itself has been reduced to a 'specious present,' and the social engineering of time into a micro-managed prism of empty granularities. Is it possible, just possible, that what Nietzsche described as ressentiment, this furious reaction formation at our own distorted instincts, now makes a new appearance at an exclusively *cultural* level? Is the real meaning of post-history the cultural road stories of a civilization suiciding itself to *virtual* life?

Just as nuclear warfare gutted history, so too genetic engineering vacates the body. Suddenly and unpredictably, a new master discourse under the triumphant sign of *biology as destiny* has installed itself as the epistemological lynchpin of a global alliance of the so-called *life* sciences and the *life* industries. Cloaking itself in the antiseptic, technical language of genetic engineering, hyping itself as a 'bible of life,' institutionalizing itself as the Human Genome Project, here promising a future medical cornucopia of gene therapy, there warning against the dysgenic effects of unmodified organisms, everywhere dreaming anew of the genetic perfectibility of the human body, the language of biology as destiny marks the appearance of what I call 'third-wave eugenics.' Having successfully immunized itself from the overt fascism of the second-wave eugenics of National Socialism and once having distanced itself from an open affiliation with its Darwinian and Mendelian origins, third-wave eugenics projects itself into the future as the spearhead of the will to technology. Nihilism today speaks the language of biology as destiny. The culture of third-wave eugenics is only awakening to its possibilities. We are, I believe, entirely unprepared for this transformation.

Culturally, it is as if we are living through the cultural trauma of two abrupt, and ethically unfathomable, shutdowns: the ending of a progressive sense of history and an indeterminate sense of time with the climactic events of WWII; and the ending of an understanding of the body as something more than its genetic code. We are the victims of *two* Manhattan projects: one resulting in the extermination of history, and the other in the cryptography of the body. Might not cultural trauma of this pervasiveness not also serve simultaneously as both a precondition for the seduction of genetic determinism and an anticipatory sign of its coming triumph? Viewed ethically, shouldn't such 'big science' as the Human Genome Project not also be considered in the psycho-ontological language of trauma: the certain outcome of a world culture that, once committed to the language of technology as destiny, now finds itself exhausted, fatigued, feeding on its own referentials, while all the while warming itself in the sun of technicity? Heidegger, Marx, and Nietzsche, then, as trauma theorists diagnosing in advance the cultural preconditions necessary for the triumph of the will to technology as well as its nihilistic fallout.

Artificial War

> I think that space, in and of itself, is going to be very quickly recognized as a fourth dimension of warfare.
>
> General Ronald R. Fogleman, USAF, Ret

Not just artificial life, but also now artificial war.

Consider, for example, the recent war in Afghanistan, where, in an epochal break with traditional military strategy, RQ-1 Predator Drones equipped with Hellfire antitank missiles were utilized *both* as stationary platforms for long-term optical surveillance and as remotely controlled missile launchers. *Real-time proximity* (surveillance of the caves of Afghanistan) combined with strategies enhancing *virtual control* (those video screens in Washington displaying action on the ground in a remote battlefield) – this technological mediation of the hyper-modern technologies of the twenty-first century with medieval tribal warfare of the third century – suddenly migrates war to the planetary, digital dimension for purposes of space-based information warfare.

With this, the age of Artificial War has begun. In its manifesto for the future of cyber-war, *Vision 2020* (www.af.mil/vision/), the newly created *United States Space Command* theorizes a future battlefield of 'full spectrum dominance.' Abandoning the earth-bound dimensions of land, sea, air, *USSPACECOM* projects a new era of artificial war in which the battlefield occurs in the 'fourth dimension' of space. Befitting a 'space-faring nation' such as the United States, third-dimensional warfare is surpassed by a vision of future war in which 'battle managers' are, in essence, computerized editing systems running on automatic, absorbing fluctuating data fields concerning attacks and responses, monitoring satellite transmissions from twenty thousand miles in deep space, sequencing missile launches, integrating 'dominant maneuvers' in space with 'precision engagement' on the ground, sea, and air, providing 'full-dimensional protection' to 'core national assets' and 'focusing logistics' for a virtual battlefield that stretches into an indefinite future. As USSPACECOM theorizes, the control of the seas in defence of commercial economic interests and the war of the western lands in defence of the expansion of the American (technological) empire to the shores of California has now migrated to a war for the 'control of space.' Consequently, a future of artificial warfare in which space itself is weaponized. *Fourth Dimensional warfare* is the technical language by which the American empire now projects itself into a future of Artificial War: a *fourth-dimensional rhetoric* of 'global engagement,' 'full-force integration,' 'global partnerships,' weaponized space stations, tracking satellites, reusable missile launchers, and on-line, real-time remotely controlled anti-missile systems.

I emphasize this story because it is revelatory of the meaning of the will to technology. Here, technology not only is the *chosen aim* of technological instrumentality (weaponizing space), but also involves *technologies of mythology* (the well-rehearsed story of the unfolding American frontier where wagon trains evolve into Predator Drones, and sea-faring navies migrate into space-bound automated battlefield manager systems), *technologies of thinking* (the fourfold 'tactics' of space war: dominant manoeuvre, precision engagement, full-dimensional protection, focused logistics), and *technologies of (aggressive) judgment* ('multinational corporations' are also listed in *Vision 2020* as potential 'enemies' of USSPACECOM).

More than futurist military doctrine for the twenty-first century, *Vision 2020* represents the essence of the will to technology. Here, technology is both a space-faring *means* to the successful prosecution of artificial warfare and its sustaining *ethical justification*. The will to technology folds back on itself – a closed and self-validating universe of thinking, willing, judging, and destining – that brooks no earthly opposition because it is a will, and nothing else. As Nietzsche reflected in advance: 'it is a will to nothingness.' Or, as Hannah Arendt eloquently argues in her last book, *The Life of the Mind*, 'the famous power of negation inherent in the Will and conceived as the motor of history (not only in Marx but also, by implication, already in Hegel) is an annihilating force that could just as well result in a process of annihilation as of Infinite Progress.'[1] Could it be that the world-historical movement captured by the military logic of *Vision 2020* – this command vision of America as the historical spearhead of the will to technology – represents that which is probably unthinkable but consequently very plausible, a contemporary expression of the metaphysics of *'not-being'*? If 'permanent annihilation' is the sustaining (military) creed of *Vision 2020*, then this also indicates that the world-historical movement, which it so powerfully strategizes, is driven onwards by the seduction of negation, another suicide note on the way to the weaponizing of space.

Consequently, if the American novelist Don DeLillo can write so eloquently in his recent essay, 'In the Ruins of the Future,' that 'technology is our fate, our truth,' this also implies that in linking its fate with the 'truth of technology,' the United States, and by implication the culture of globalization, may have, however inadvertently, infected its deepest political logic with the will to nihilism. In the sometimes utopian, always militaristic, language of technological experimentalism, 'Not-being' finally becomes a world-historical project. Those who are only passive bystanders to the unfolding destiny of the contemporary American descendants of the Puritan founders can only look on with amazement coupled with distress as the 'American project' embraces not only the weaponizing of space but also genetic experimentation with the question of evolution itself. While DeLillo goes on to say that technology 'is what we mean when we call ourselves a superpower,'[2] his pragmatism sells short the point he really wants to make: namely, that by linking its fate, its truth, with the question of technology the United States has also enduringly *enucleated* itself within the larger historical, indeed if *USSPACECOM* is to be believed, *post-historical, project of technology*. Enucleated not as something other than the technological destiny which is its profession of faith, of truth, but enucleated in the more classical sense of the term, of being somehow interior to the unfolding destiny of the will to technology. The larger cultural consequence of this bold act of willing remains deeply enigmatic. In this case, is the will to technology an intensification of the pragmatic spirit upon which the American experiment was founded? Or has the will to technology, at the very moment of its historical self-realization, *already reversed its course*, becoming its own negation: Arendt's prophecy of 'not-being' as a 'process of annihilation'? On the ultimate resolution of this question depends the American fate, the American truth, as the spearhead of technology.

On the public evidence, what makes the American project truly distinct today is its enthusiastic abandonment of the pragmatic will for the uncharted metaphysical territory of 'not-being.' The will to the conquest of empty spatialization and the vivisectioning of the code of life itself has about it the negative energy of suicidal nihilism. Here, the language of 'not-being' – the desiccating logic of what Heidegger memorably termed 'Nothingness nothings' as the historical form of the technological project of 'permanent annihilation' – expresses itself vividly in two master commands: *Space Command* and *Genetic Command*. The first operates in the language of *weaponized astrophysics* where the curvature of space is manipulated for strategic purposes, and the other *sequences* the human genetic code itself. *Thus, control of space is inextricably linked with control of time.* The dynamic will to technology projects itself doubly in the macrophysics of a 'space-faring nation' and the microphysics of a body-faring cellular biology. This is a collective demonstration of hubris that Greeks in the classical age would only admire, and then fear, for its (technical) audacity and stunning (metaphysical) innocence.

Ironically, at the very instance that *USSPACECOM* projects an imperialist military future of 'full-spectrum dominance,' 9/11 occurs and we are suddenly time-shifted into the age of viral terrorism. Similar to the incommensurability of technology itself where the reality of 'permanent annihilation' is sometimes offset by other ways of *thinking* technology, the human imagination does not begin, cannot begin, with tactics of 'dominant manoeuvre' and 'precision engagement' and 'full-dimensional protection' and 'focused logistics,' but with the terrorist side of fluid, earth-bound, real material warfare.

Artificial war, then, as a prolegomenon to the codes of technology.

2 Streamed Capitalism, Cynical Data, and Hyper-Nihilism

Theory is sometimes read best as an art of anamorphosis: an aesthetics of distorted perspectives and obliquely mirrored angles of vision that when positioned correctly illuminates and clarifies the controlling logic of culture and society. That is why in the age of completed nihilism when the will to technology strives to achieve apogee with the final conquest of time and space, it is salutary to revert to the untimely meditations of the true prophets of technological destining: Heidegger, Nietzsche, and Marx. In the intensity of their insights as well as in their implacable demand to understand the deeper civilizational crisis precipitated by the epoch of cynical data, streamed capitalism, and hyper-nihilism – to understand, that is, the *destiny of technology* – their thought effectively mirrors the closed, self-validating rhetoric of the present.

Against the positivism of technological rationality, Heidegger insists on grasping the essential element of metaphysics that lies unexposed and unheard in the technological maelstrom. In his thought, the 'question of technology' as not-being is privileged as a primary clue to understanding the trace of nihilism the origins of which lie in the split consciousness of techne and poiesis, these doubled stars of the technological dynamo. The primary opponent of his own thought, Nietzsche's insight into the disappearance of power into a 'perspectival simulacrum' gives the lie to the comforting rhetoric of the 'will to power,' preparing instead for a more radical interpretation of the 'will to will' as the essence of the question of technology. Impatient with the temporal duration of history, Nietzsche reverse engineers the flow of contemporary power until its surface begins to surrender concept-phantoms from its past *and* future: the maggot man, cynical power, suicidal nihilism, the ascetic priest, beings half-flesh/half-ressentiment. The same with Marx's *Capital*. This ideological gravedigger from the past refuses its rites of interment, throws off its historical association with early capitalism, and begins to theorize anew, insisting that *Capital* is a critical philosophy whose self-realization lies in the future, not the past. Here, *Capital* reverse engineers the contemporary era of streamed capitalism, theorizing anew the destabilization of the commodity-form as it is thrown into an indefinite cycle of smooth circulation, as exchange-value mutates into the knowledge-theory of value, and as capitalism itself migrates from the theatre of representation to the order of (financial) virtuality.

Could it be, in a sublime twist of intellectual history, that theory is always only truly critical after the fact, that the flow of granulated time itself sometimes deepens thought which, once removed from its immediate historical circumstance, is finally liberated to be the *intensity* that it always sought: a process of pure conceptual undermining cut with the bitter aftertaste of philosophical futurism from the grave? We, the first citizens of the twenty-first century, are privileged witnesses to the empire of (technological) spatialization as it extends its ocular coordinates into deep space and deep flesh. Nietzsche might have spoken of 'vivisectionist thought' but it is our curious fate to live in a culture of almost surgical-like deconstructions of the social body. Heidegger intimated that 'profound boredom' would be the dominant sign of completed nihilism, but, in this epoch, boredom falls upward from concept into the daily experience of the distractions of speed culture and the streamed emptiness of the new bourgeois ego. Marx wrote out a prolegomenon to the age of the circulating commodity, but it is our peculiar technological destiny to experience the speed of circulation as the image matrix within the ecstatic signs of which we are only humiliated flesh.

Marx, Heidegger, and Nietzsche provide a comprehensive, critical, and futurist account of the destiny of technology: its origins, implications, and historical method of realization. Marx, the theorist of the political economy of technology, relates in precise detail how the industrial stage of capitalism will give rise to capitalism as a 'pure circuit of circulation' – *streamed capitalism*. Nietzsche, the poet of technology, diagnoses a century in advance the invidious growth of *nihilism* as the cultural sign of technological society. Heidegger, the metaphysician of technology, deepens our understanding of the mythic origins of technology to include a critical reflection on the *meaning of technicity* in relationship to the more fundamental question of being. While Marx privileges an understanding of technology in terms of the question of capitalism, and Nietzsche dwells on the fate of subjectivity in the society of fully realized technicity, Heidegger's lasting contribution is to be a bridge between Nietzsche and Marx, unifying Marx's account of the axiomatic of virtual capitalism with Nietzsche's genealogy of digital morals. For Heidegger, understanding 'technology as destiny' necessarily calls into question conflicting impulses in the digital future towards past and present: technology as promise and danger. In this case, while Marx 'presences' the question of technology in a historically specific account of capitalism as the dominant historical representation of technology, and Nietzsche 'poets' technology against the background of the more ancient impulse in western civilization towards nihilism, Heidegger provides us with a chilling, complex, and internally nuanced meditation on the general metaphysical crisis in contemporary culture unleashed by the triumph of the will to technology. Consequently, while Heidegger teases out the nihilism in Marx's description of the commodity-form, and Marx materializes Heidegger's vision of technological destining in the spectre of surplus-exchange, the alienation of labour, and the fetishism of the virtual commodity, Nietzsche *poets* the question of the will until it drops its mask of instrumentalism and assumes the sign of nihilism as its future-drive.

More than is customary, Marx, Nietzsche, and Heidegger are theorists of the future. About this, Nietzsche was explicit, declaring, for example, in *The Will to Power*, that he was writing prophecy for the next two hundred years. Marx, the theoretician of the gathering storm of class struggle against bourgeois capitalism, wrote projectively, his entire theoretical discourse a reflection that would not be complete until the capitalist axiomatic had actually run its course. And Heidegger? While fully attentive to the present development of technology in the doubled languages of technicity and biogenetics, he insisted that the question of technology always folded back past and present on the future. In his vision, understanding technology was always coeval with meditating on the question of the genealogy of destiny itself, brushing the language of technical calculation against poetic utterance in order to produce a general theory of the crisis of civilization, a more proximate account of the meaning of technicity in terms of its mythic origins and digital future. For Heidegger, our future is already framed by what has been most forgotten – 'the oblivion of being.'

The individual inflections brought by each of these three thinkers to the question of technology are strikingly different. Marx interprets the rise of the commodity-form through the optic of a political analysis of class struggle. Nietzsche is the writer of hermeticism, tracing a tortured pilgrim's progress into the deepest recesses of his own subjectivity as the privileged lens for revealing the character of passive and suicidal nihilism. Heidegger is ultimately a cosmological thinker, measuring the fate of increasingly technical subjects against a broader and more abiding reflection on the metaphysics of being as it remains shrouded under the haze of the forgetfulness of the gods. In effect, three deeply incommensurable thinkers, simultaneously conflicted in terms of their most essential theoretical attitudes and, in the deepest sense of the term, their political outlooks. And yet, it is precisely the impossibility, this incommensurability, of the strange thread of Marx, Heidegger, and Nietzsche that makes of their analysis a synthetic account of the will to technology. If technology itself is an incommensurable experience, drawing together opposing tendencies towards instrumentality and creativity – technotopian visions of the global brain and virtual class perspectives on the maximization of share value – then the only effective strategy for understanding the broader implications of the will to technology is to theorize at the edge of incommensurability: forcing the will to technology to reveal its secret destiny by framing the question of technology in the language of impossibility. And that's what the conjunction of Marx, Nietzsche, and Heidegger is: a theoretical impossibility in an incommensurable digital experience. So then, an impossible theory for an incommensurable experience.

While the interpretive strategy of stressing Marx, Heidegger, and Nietzsche, insisting that their writings be thought with and against the practical vicissitudes of technological experience, illuminates the essence of their theories – *hyper-materialism* (Marx), *hyper-metaphysics* (Heidegger), *hyper-nihilism* (Nietzsche) – it should also be said that they can be so productively *hypered* because of the deep affinities of what they have to say. Set in the actual historical context of their writings – Marx the theorist of the fateful labour insurrections and capitalist resistances of the nineteenth century, Nietzsche the philosopher of approaching ruins of the twentieth century, Heidegger, the metaphysician, projecting mid-twentieth-century experience back to its sources, back to the ancients, and forward to the post-human – the triad of Marx, Heidegger, and Nietzsche remains strangely isolated, what they have in common quickly extinguished by the necessary, particular oppositions among politics, metaphysics, and nihilism. However, now that we live in the twenty-first century, a century already deeply defined by the triumph of liquid flows of globalization and the practical defeat of (local) political economies, by profound forgetfulness of the ethical ambiguities of the technological dynamo and by the elimination of 'thoughtfulness' itself in public and private life, and by the acceleration of Nietzsche's sense of passive and private nihilism in the hybrid form of 'hyper-nihilism,' the essential commonalities of Marx, Heidegger, and Nietzsche as visionaries of the present crisis finally allow them to be thought together. Not brought together in terms of an artificial unity of intellectual history, but synthesized in the important sense that their writings constitute the ruling code of the will to technology. Marx, Heidegger, and Nietzsche as the three defining moments of the will to technology. Indeed, if the twenty-first century will be lived out under the sign of the will to technology, then the fate of Marx, Heidegger, and Nietzsche is to be the litmus of the future: benchmarks against which the will to technology continues to test itself. In that curiously doubled sense of technological life where every technical innovation is simultaneously a creative probe and an erasure, the real importance of thinking Marx, Heidegger, and Nietzsche is that they are the double probe thrown up by the will to technology, simultaneously its fiercest critics and, most certainly against their own intentions, accurate guides to the unfolding technological future.

For example, an intensive understanding of the unfettered flows of virtual capitalism can only begin with Marx. His thought is explicitly how the will to technology materially realizes itself, speaking quietly in the pages of *Capital* of a coming future that will be marked by the acceleration of the *form* of capital itself into a pure virtual form. Heidegger writes out the essence of technology. Nietzsche poets nihilism. Might it be possible that by a curious gesture the will to technology, drunk with hubris, fully confident of its sovereignty, is itself gambling with the gods of fate – allowing its deepest tendencies to be crystallized, confident of the death of thought? Or is it something more: could it be that Marx, Heidegger, and Nietzsche are transformation matrices in the will to technology, fatal points of energy and stasis where opposite tendencies meet, and that what is really being worked out in their texts is our own personal futures? That is why these are thinkers of impossibility. They have become the history that was the object of their theorizing. Their thinking is the distilled essence of the codes of technology: *Nietzsche the genealogist of cynical data; Marx the prophet of streamed capitalism; and Heidegger the radical metaphysician of hyper-nihilism.*

Recombinant Capitalism

Consequently, the real importance of Marx is that his thought is equivalent to the era of fully realized capitalism. In Marx, the commodity-form transcends the model of production in the direction of virtuality. The labour theory of value gives way to the knowledge theory of value. The class warfare of bourgeoisie and proletariat based on the antagonistic relations of the exploitation of surplus labour-value is transformed into a newly imminent global class antagonism of virtual class and surplus class based on the extractive values of capitalism as a flesh machine. Here, the commodity-form abandons the dialectic of exchange and production, splitting the difference as digitality. Approaching capitalism critically and with an eye tuned both to the revelation of capitalism as despotic power and to the (proletarian) possibility of overcoming capital, Marx's final contribution was to theorize the legacy codes of the new capitalist order: *virtual capitalism*. Captured by the fatal spell of the dialectic, capitalism itself is Marxism recombinant.

Capitalism as Marxism recombinant? That means that Marxism today has accelerated to such a point of delirious intensity that capitalism itself comes under the spell of Marx's vision of dialectical materialism. From the grave, Marx brilliantly framed the future of virtual capitalism: its motor-force – the digital commodity-form; its theory of exploitation – the knowledge theory of value; its class struggle – the virtual class versus the surplus class; its key vision – the speeding up of the model of production to the point that it disappears into the spectre of virtual commodities.

With Marx, the secret of virtual capitalism is first revealed: the movement from the commodity-form based on a labour-theory of value to the virtual commodity-form, the rise of the knowledge-theory of value, surplus-labour, surplus exploitation, surplus-alienation. Marx is simultaneously the decoder of the surplus positivity of capitalism and its negative moment: the rise of the anti-virtual class, the splitting of the knowledge-class into two irreconcilable parts, the defeat of particular labour history is matched by the emergence of a universal labour movement. The virtual class is entangled in the dialectic of Marx's thought. Capitalism thought it had defeated Marx with the political defeat of official socialism in the east. This is an illusion. Liberated from its subordination to state ideology, Marxism is finally free to be fluid, liquid, deterritorialized, a nomadic theory of alienation which does not grovel to state forms. While defeating Marx is the decisive challenge of virtual capitalism, this cannot be done without capitalism overcoming its own strategy of surplus-exploitation, or without transcending the speed of circulation of the circulating commodity-form. However, to do this – to enter the stage of fully realized capitalism – capitalism would have to overcome itself. Trapped in the mirror of Marx's thought, capitalism will either be completed by the Marxian theory of the circulating value-form of the digital commodity, or it will endlessly recycle in the loop of incomplete commodification. Streamed Marx is the future of capitalism in its mature phase of full self-realization.

Streamed Capitalism

Marx's theory of the commodity-form, finally released from its staging area in political economy, projects itself into the future as the trace of virtuality present in speed capitalism. When Marx's *Capital* is upgraded by digitality, the commodity-form finally accelerates beyond the laws of the old economy, throws off the nostalgia surrounding class relations determined by the socialization of production, and goes over to the consumption of virtuality. In uneasy terms, the commodity-form lets us know that it was never all that invested in the material outcomes of the process of capitalist production, that it never really wanted to come out of the cycle of circulation, that capital never rested easy with its specific historical representations in the myth of production. Emancipated by the termination of the link between Marxist theory and the now debunked state project of official socialism, Marx is emancipated to be understood for what he always was: the preparatory theorist of virtual capital. Read *Capital*, then, not as a now historically superseded theory of industrial capitalism, but as a profoundly insightful vision of the disappearance of capital itself into virtuality. Today, just because the theory of 'dead labour' has been replaced by the irreality of 'dead capital,' Marx can triumphantly return as the first and best of the digital futurists. What Heidegger describes as the culture of 'completed metaphysics,' and what Nietzsche laments as the 'will to power,' Marx patiently deconstructs. The theorist par excellence of the legacy codes of capitalism, Marx first grasps the axiomatic of virtual capitalism: the disappearance of the circulating commodity into the pure speed of circulation; the transformation of labour from a 'factor of production' to the production of 'factored labour'; the ascendancy of the 'knowledge' theory of value; the rise of the virtual class as the historical objectification of 'value valorizing itself'; and the relentless disciplining of capital by the will to technicity. In Marx's vision of the dynamic transformation of capitalism from a theatre of representation to virtuality, the new capitalist axiomatic stands revealed as the mirror of velocity.

A metaphysician most of all, Marx wrote about the technology of harvesting labouring bodies, when the alienation of labour was already in the process of being replaced by the virtual labour of alienation, where the capitalist axiomatic refused its necessary connection to the logic of accumulation, going over to the virtual side of radical dis-accumulation. When Heidegger rehearsed in his writings the contemporary fate of subjectivity as 'being held out into the void,' he might well have been reflecting on the metaphysics of contemporary capitalism. Stripped of its illusions about the necessity of capitalist accumulation, turning its back on the stolid model of production, substituting the digitalization of dead knowledge for living labour, Marx's *Capital* is the essential nihilist. Held out into the void, simultaneously its own goal and the guarantee of its own stability, Capital is 'weakened in its ground.' It has no necessary aim since its only necessity is to be aimless. It has no essential ground because its destiny is to instrumentalize the question of essence, to mutate materialism itself into a process of increasingly virtual means. It can never really be secure about being the guarantee of its own stability because, as a process of instrumental means, its only guarantee of survival is to continuously destabilize itself. Consequently, the fate of Capital is to turn on itself, to transform the once-hoarded contents of capital accumulation into the excremental representations of the disappearance of capital. Sick of itself, Capital wants to be rid of itself. It has become weary of the referential weight of products. It desperately desires to be post-human, to liberate itself from the regime of the signs of capitalist accumulation in favour of capitalism as a regime of virtualites. It wants to escape the weight of time, to illuminate its future with the lightness of virtual exchange. Growing increasingly resentful, Capital turns inward. It is meditative in its desolation. It turns darkly. It turns viciously. It is the animal of economic cruelty. It drifts. It eats. It incorporates. It vomits. It calls its spew advertisements. Here, it dresses up properly to be put on stage-managed displays as a model of economic productivity for the commercially agnostic. There, it falls sideways and upward and often even downward into its own emptiness. It is the quintessential expression of Heidegger's concept of the 'emptiness that bores.' Capital is the dominant historical expression of contemporary nihilism.

Heidegger is necessary to the project of overcoming Marx, incorporating Marx's historically specific analysis of the commodity-form of capitalist production into a virtual analysis of the productive mode of the digital commodity-form. Perhaps it has always been this way. Maybe in the ironic game of thought the lasting contribution of Heidegger was always to be this: to be present at the stage of completed nihilism in order to disclose Marx's *Capital* for what it was. Not simply a theorization of capitalism as a definite form of political economy, but of something much more ominous: capitalism as the essence of nihilism. In this sense, Heidegger is the truth-sayer of Marx as sorcerer-priest. In the same way that Heidegger would often repeat that technology cannot be understood technologically, so too Capital is not understandable exclusively within the terms of political economy, but only as the latest expression of a much more mythic drive in the human condition. Call it what you will, the 'geist' of history, the drive to the 'homogenous unitary state,' the 'acquisitive' will, or perhaps the 'will to technology,' capital is always a derivative historical form. It is metaphysics masquerading as political economy. *Capital is the metaphysics of the not-will, a will to exhumation masking itself in the alibi of production.*

Consequently, Marx can draw out so vividly the speed of circulation at the centre of hyper-capitalism because Marx's thought is itself the essence of technological consciousness. Here, Marx's thought analyses the historical laws of movement of the commodity-form with such relentlessness, hovers around the 'mystery' of commodity-fetishism with such intensity, can be so spellbound by the speed of the cycle of circulation, that Marx himself becomes the leading prognosticator of hyper-capitalism. A prophet without a determinate country, a futurist of the past, a theorist of political economy with such creative intensity that the 'object' of political economy – the dialectical materialism underlying the system of capitalism – dissolves before his analysis, revealing only the absent, indifferent movements of late capitalism which, having disappeared into the cycle of circulation, refuses all production in favour of virtuality. Ironically, while Marx possessed full critical consciousness of the material history of capitalism, he was necessarily forgetful of his own intellectual legacy as the first and best prophet of the transformation of capitalism into technicity. That is, of course, the Heidegger in Marx: that point where the capitalist axiomatic transcends the limitations of political economy, finally making its appearance as the material history of completed nihilism.

Cynical Data

Nietzsche speaks out of the unfolding skin of the will to truth, that point where the truth breaks its conventicle with Christian morality

and prepares to travel alone in the radical individualist style of technological rationality. Born with the prophecy of the death of god implanted in his nervous system, Nietzsche walked the outer planets of
human solitude, stopping here to write the pilgrim's tale of *Zarathustra*,
there to speak contemptuously of the *Twilight of the Idols*, drinking
deeply at the fountain of the will to anti-truth. Tattooed by his writing,
Nietzsche is the will to power made flesh in the body of this speed
philosopher of the deep loneliness of human solitude. Like one of those
bleak-head thinkers who opened his eyes to the danger that he thought
he was always dreaming, Nietzsche was split consciousness for a radically divided history. Half-narcissist/half-prophet, he had that kind of
mind-nerve that felt compelled to think itself into the difference that
was the disappearance of its own being: a fully incommensurable philosopher. A strange space of difference, a contradiction, a restlessness, a
going across, a going away, a going down: the name of Nietzsche is the
future of the past that was the present of his despair.

Maybe that is why the name of Nietzsche continues to haunt the new
century. Like one of those seduction amulets on the wall of a dusty,
broken-down bazaar in the dark-thought zone, Nietzsche projects you
instantly into the 'spider web' of nihilism. Eyes most of all are always
sightless about these things. But one day the sun rises on the shutting
down of the eye, the censor of ocular perception is dropped, and the
voice of Nietzsche begins to be heard with the ear. It's a strident,
polemical, but strangely lyrical voice, and it tells the story of a mind
cutting right through the nervous ganglia of culture once and future to
its core pain. Travelling deep in the genealogy of morals and riding
high on the peaks of the will to truth, it speaks in fast aphorisms and
slow poetics in its meditations of the blood. And it has got a voice
jammed tight with perspectives insightful, and prophecies maddening,
and metaphysics unthinkable, and psychological insights unbeatable.
And once that voice speaks, once it speed-writes *Thus Spake Zarathustra*
and *Twilight of the Idols* and *On the Genealogy of Morals*, it is a voice that
cannot and does not stop. It is fuelled by resentments deep against
Wagner, against the Germans, against the priests, against the 'maggot
man' and against the 'blond beasts,' but that is all just an emotional
stepping-stone. Because Nietzsche's is a voice in the flesh spirals of the
inner ear that rants its way into a withering vivisection of everything
that comes within the orbit of its madhouse boom. It is hungry for
objects. 'Apparent history' and the 'will to truth' and the 'will to power'
and 'ascetic priests' and 'mechanical forgetfulness' and the 'will to will'

and 'suicidal' and 'passive' nihilists and the 'last man': it just fast-speaks its way into the vector veins of the future. And so, you're reading Nietzsche and you're speaking Nietzsche and you're thinking Nietzsche, and Nietzsche is speaking and reading right back, and you know that you are in a crazy gig whirlwind, that that voice is taking you to *Alice in Wonderland* places maybe you shouldn't go. But the looking-glass is pretty and the will to truth is transparent, and so you step through the looking-glass of Nietzsche's mind. Stepping through is easy. Returning is another matter. It's like taking a full-barbed meta-physical hook out of the distended eye or deflating the ramped-up ear or making dormant again those thoughts hip and bleak that take you on a direct-acid trip to the culture zone.

Nietzsche as the will to truth? Perhaps this dissembling nomad of the nocturnal sea winds was putting into practice the perspectival simulacra that he thought he was only writing. Maybe he was a prophet-machine installing the reality of the insights that he thought he was only imagin-ing. If this is so, could it be that Nietzsche never died? That he's the despotic spearhead of the will to truth that is finished now with cyber-sex and cyber-war and cyber-bodies, finished because *the will to truth is its own overcoming*? Perhaps the spirit of the will to truth as its own overcoming is the real essence of the Human Genome Project with its predictions fantastic of artificial life and the death of death itself. In the hubris of biotechnology the will to truth granulates itself in the molecu-lar matter of genetics. When the project of the will to truth is embedded in the language of genetics, then truth itself can be abandoned because knowledge now is understandable only in the language of life. The Human Genome Project, then, as Nietzsche's last aphorism. In a strange simulacrum of virtual cryogenics, the name of Nietzsche, with its furi-ous 'uncovering' of the nihilism of the will to truth, lives on in reverse form as the animating spark of the will to anti-truth that is the essence of contemporary techno-culture. Nietzsche, then, as a gravedigger of the death of god, but the midwife of the birth of post-human. The radical will to negation as the surplus-positivity necessary for the functioning of a system that runs, and can only run, on its own excremental residues.

For a philosopher who proclaimed himself prophetic of the next two hundred years, Nietzsche's thought always travelled backwards: a scream of philosophy that jammed its way into the future by focusing its meditation on the twilight of culture that would appear with the death of the Christian god. In the epic tradition of myth, Nietzsche appears now as the last messenger of dead gods. An uninvited guest at the digital feast, he enters the banquet hall of technological instrumentalism with a message and a warning. The message is expected. God is dead. The warning is welcomed. Nihilism raps on the door. The transcendent has lost its appeal. The twilight of the idols has taken possession of human flesh.

But we know different. God has been virtualized. We have already simulated his absence. We are the most theological of peoples. We are preparing for the Apocalypse. We are eager for the resurrection-effect. We know all about the death of truth because we're the most moral of people. God may have died in our souls, but not in the wires. There are many digital gods circulating in Net space. Many virtual modes of transcendence: a noosphere of externalized ocular perception for the gods of speed; netted bodies labouring in the disciplinary factories of the mind for the gods of work; image flesh digitized and edited and reconfigured for the gods of simulation. We are global gods. We are technicians of the soul. We are vivisectionists of the codes.

Nietzsche's legacy was to reinstall the principle of transcendence that he thought he was finally negating. Not the anti-Christ, but the energizing suicide machine of nihilism: the death of the referential signs of the social. In techno-culture, we live on the negative energy of the (digital) emptiness within: the downloading of culture and society and sex and consciousness and economy in the waiting data vats. We are energized by the anti-matter of disappearances. We are an aesthetics of extinction, a diagnosis of stripping clean, a navigator of the differences, and designer-machines of wild negation. We have learned much from Nietzsche. He can be the sign of 'completed nihilism' because his thought is the fundamental 'turning' of power into its opposite: the decline of life into the will to virtuality. He announces the nihilism that makes possible the emancipation of cynical data as the ruling code of the technodrome. The shadow of Nietzsche is everywhere. Digital being as a 'going under.' The global brain as a 'crossing-over.' Flash effects as an 'adventure,' a 'gamble,' a metaphysics of special effects. Connectivity as a 'danger' and a 'possibility.' Nietzsche is the Christ of anti-truth as the animating energy of the technological: the resurrection-effect of an axiomatic of anti-truth that works only to confirm the impossibility of slipping away from the will to truth. Nietzsche is the end and the beginning: the final jest of the ancient gods, still grumbling over the despotism of the Christian god. Nietzsche is the anti-Christ of digital being.

Hyper-Nihilism

Tom King. *Untitled*, 2002.

> If, finally, metaphysics is the historical ground of the world history that is
> being determined by Europe and the West, then that world history is, in an
> entirely different sense, nihilistic.
>
> Martin Heidegger, 'The Word of Nietzsche'

There is a painting by a young Toronto artist, Tom King, which perfectly expresses the culture of boredom. A painting without a name but
with a definite attitude, the image is of an androgynous, almost hybrid,
male, muscular and naked, with short-cropped hair and an intense
stare of blank fury that pushes its way out of the canvas to take posses-

sion of the surrounding space. Here the body relaxes in a kind of forward lounging gesture, a crackling line of energy, which insists that this body will not have its presence denied.

Curiously, as if to erase the hostility of the eye, the painter has reworked the original composition by hand, smearing the paint so as to make anonymous the subject of the figurative pose. A stare without a name. A body without a geography. Figuration without descent. A void face with a menacing stance.

It is precisely in that empty space between the entirely personalized, entirely hostile, intensity of the stare and the physical disappearance of the identifying lines of bodily figuration that the painting gains its unique aesthetic, and emotional, power. In that space of difference we are suddenly taken into the space of an annihilating emptiness: what Heidegger once described as 'being held out into the void.' An attitude of vacancy surrounds the painting: this image of the anonymous, smeared male figure may have been delivered up from the history of art, but what appears is the slightest inflection of undecidability. That there is nothing left to decide is the aesthetic sign of the painting. Here, we are present at the birth of the image of the body in the time of the *suspended now*, posed and waiting and folding back into itself – *vacant being* that is simultaneously alert and drifting. A painterly gesture signifying the coming into presence of a culture of 'profound boredom.' An image of being as boredom that bleeds off the canvas directly into the philosophical meditations of Martin Heidegger.

For Heidegger, 'profound boredom' is the fundamental attunement of the era of completed nihilism. Marx might have prophesied the death of the social as the inevitable historical legacy of the will to capitalist accumulation. Nietzsche might have meditated deeply upon the will to power – cynical power – in the shadow of the death of god. However, standing apart from these searing visions of the death of the social and the death of god, it was Heidegger's special insight to understand that the future coded in the language of technology spoke quietly, but no less insistently, of another death – the hyperbolic sign of the death of the human. A fabulist no less than a metaphysician, Heidegger understood that the famous 'question of technology' nested the advent of another mythology – the story of the bored will which, once released into history, would return to consume its human origins. Never capable of adapting to the utopian pretensions of the vernacular of the 'post-human' and certainly never willing to subordinate his thought to the 'enframing' necessary for setting in motion the regime of completed technicity, Heidegger grasped into his thinking the full implications of the growing boredom of the will. For Heidegger, we, the inhabitants of late modernity, are pilgrims in an entirely new metaphysical phase of western (cultural) history. If this is an age of 'completed nihilism' with Nietzsche as its prophet and unwitting intellectual agent, it is because the unfolding story of the will – a movement that has previously coded economy (the will to accumulation), politics (the will to power), religion (the will to believe) and culture (the will to seduction) – has already undergone a radical transformation. In Heidegger's theoretical imaginaire, it is not simply that the language of technology is the dominant metaphysical impulse of culture and society, but that in this primal impulse there resides a 'twofold essence': an incommensurable moment of revelation and oblivion.

Embracing the question of technology as itself an indispensable language of revelation, Heidegger was apocalyptic in his prognosis of what is disclosed by the coming to presence of completed technicity: *'objectification'* as the result of technological willing; the *'harvesting'* of humans, animals, plants, and earth into a passive 'standing-reserve' waiting to be mobilized by the technical apparatus; the liquidation of subjectivity itself into an *'objectless object'* streamed by the information matrix; the appearance of *'profound boredom'* as the essential attunement of the epoch of the post-human; the potentially fatal transition of the language of destining to the last *'enframing'* – post-human culture under the ascendant sign of 'completed nihilism.'

In Heidegger's political vision, the metaphysics of terror explicit in Christianity now inscribes itself in the cybernetics of the (technologically) possessed body. Simultaneously its own ground and goal, power in the era of 'completed nihilism' roams the territory of the post-human – its 'harvested' bodies, 'mobilized' minds, 'corrected' animal spirits, and 'objectless' desires – in the circulatory language of the 'will to will.' Folding back on itself in a violent, circulating loop of harvesting and projection, the will to will functions as a pure relation of force. It is a *technological* will because it projects the instrumental language of technicity as the dominant metaphysics of the culture of the post-human – the codes of objectification, mobilization, the 'injurious neglect of the thing,' and the politics of the 'standing-reserve.' It is a *bored* will because, forgetful of its origins in the ressentiment of the Christian god and silent in the face of the human demand for recognition, the will to will injects the poisons of 'undecidability' and 'interminability' into the post-human condition. Stripped of its religious justifications, in-

creasingly virtualized beyond the merely acquisitive aims of capitalism, bored with its previous role as a 'condition of possibility' for the
operation of culture, society, and politics, the will to will discloses itself
in the rhetoric of hyper-nihilism. Here the will to will burns itself into
the flesh of the post-human century by three violent strategies: its
metaphysics: the transformation of the 'question of technology' into the
informing logic of the twenty-first century; its *(technological) epistemology*: the application of the disciplines of molecular biology and genetic
engineering to the streaming of natural and human life-forms into
increasingly delirious experiments in the design of the post-human;
and its *operational principle*: the circulating logic of hyper-capitalism.

Born late in the now superseded epoch of the human, we are perhaps
initiates to the more ancient games of nihilism. But not Heidegger.
Writing with complete attentiveness to the history of metaphysics,
Heidegger made the singular claim that the history of western culture
is simultaneous with the genealogy of nihilism. If Heidegger could
remark that 'Nothingness Nothings' or respond to the question 'What
is happening to being?' with the enigmatic insight that 'Nothing is
happening to being,' it is because his thought brings to the study of
nihilism the language of paradox, ambiguity, and incompossibility.
Consequently, he could analyse so darkly the grammatical codes of
completed technicity as it dissolves subjectivity into an 'objectless object' because his thought always embraced an imminently oppositional
reflex. In that strangest of ironies, Heidegger could think so evocatively
of technological nihilism because in the darkness of the post-human
century which stretched out before him his thought had about it the
confidence of a wild gamble on the riddle of metaphysics. If the question of technology presenced itself as the doubled language of 'harvesting' and 'injurious neglect of the thing,' it also foreshadowed the coming out of oblivion of the 'truth' of being. Again and again, Heidegger
returned to the fabulist theme that that which lies most concealed in the
coming to be of completed nihilism is already ready at hand, namely
the fourfold essence of earth and sky, mortals and divinity. The
civilizational crisis precipitated by completed technicity is itself a primary, and necessary, condition of possibility for the appearance of a
form of nihilism as nothingness which bears in its essential features an
uncanny resemblance to eastern mysticism and the cosmology of the
four ages. Heidegger is the completed western thinker who, breathing
in deeply the language of hyper-rationalism so indispensable to the
western myth of technology, turns the question of nihilism from the

inside by inserting a fatal element of the undecidability of being itself into the movement of the cynical will. Heidegger's thought migrates to the east as a way of seducing the bored will with another fatal story: the eastern myth of nothingness as plenitude, of the void as coming into presence, and boredom itself as an opening onto the revelation of completed subjectivity. In Heidegger's practice of the life of the mind, nihilism as 'not-being' mutates into the more ambiguous and fluid state of hyper-nihilism where technicity itself is made to drink at the fountainhead of nothingness as the certain sign of its own 'turning.' With Heidegger's vision of hyper-nihilism, the history of nihilism which we thought had found its moment of completion in the desolate technical fire-storms of data flesh suddenly stumbles, unsure of itself, turning back on its own will to will only to discover therein a way of thinking 'nothingness nothings' that makes of the supposed age of 'completed nihilism' an unanticipated preparation for a new form of nihilism: the spectre of hyper-nihilism as the turning by which the 'symbolic exchange' of the east haunts the western terrorism of cynical data.

3 Codes of Technology

The will to technology is the animating energy of twenty-first century politics and culture. Here breaking beyond the fetters of the nation-state, there dissolving the machinery of industrial production into virtual capitalism, now networking the body with the digital eye in the form of an enhanced flesh-interface, in the future transforming knowledge itself into a means of self-justification and self-interpretation for the technological enterprise, in the past preparing the way for the present reign of pure technicity, always coding, distributing, upgrading, incorporating, the will to technology is simultaneously the emblematic destiny of post-human culture and its most enduring form of self-understanding and self-expression.

The visible signs of the will to technology are everywhere. Set in motion by the software of information technology, volatilized by digital capitalism, driven onwards by the seduction of communication, and relentlessly aimed at reducing existence to the moral axiomatic of fully realized technicity, the will to technology is the sustaining myth of contemporary culture. Simultaneously an identity, a goal, an interpretation, a power, a desire, a sex, the will to technology animates the cultural noosphere. Not a will in the traditional sense of pure instrumentality, the will to technology evokes such deep fascination because it is a will that folds back on itself, sometimes projecting itself forward as pure technicity, at other times disappearing itself in the form of virtuality. It is creation. It is crash. It is consolidation. It is erasure. It is software. It is wetware. It is a moving forward. It is regression. It is a (mathematical) binary. It is (human) poetry. Both hypermodern and primitive, it can speak the language of cyberwar and viral terrorism.

Always a double movement, the will to technology is a data cannibal feeding on itself, simultaneously disappearing the actual referents of society – knowledge, sex, power, economy, politics – into nodes on the circuit of electronic production and furiously throwing itself into the future as digital destiny. A *hyper-religion*, the will to technology requires an act of faith in the efficacy of technology itself as the ritual of admission into its axiomatic procedures. A *hyper-ideology*, the will to technology is historically realized in the material form of the triumph of the virtual class. A *hyper-science*, the will to technology transforms the scientific imagination into a database for harvesting the residues of human and non-human nature. A *hyper-myth*, the will to technology presents itself in the ancient language of the gods, speaking in the more enduring terms of destiny. A *hyper-eugenics*, the will to technology ushers in a biotech future written under the sign of transgenic determinism.

Transgenic determinism? That's the dominant tendency in global cultural politics. After a long sleep during the interregnum years of the Cold War, the language of eugenics stirs again: here articulating itself in the vivisectionist visions of genetic experimentation; there pushed forward by a newly emergent form of transgenic capitalism intent on coding, classifying, manipulating, and harvesting the genetic history of humanity, animals, and plants; now working in the laboratory procedures of stem cell research, clonal propagation, gene sequencing, organ growing, and tissue replacement; later expressing itself in the conjunction of bio-sociality and artificial intelligence to produce artificial life-forms; never acknowledging its historical precedents in the first-wave eugenics of the hygiene movement of the early twentieth century or the political fascism of National Socialism; always masquerading in the cloak of a science of completed genetics.

Immunized from its historical genealogy, cleansed of its political history, speaking the scientific language of molecular biology and the economic language of the 'life industries,' transgenic determinism limits for now its discourse to the preservation of 'life' and the genetic improvement of 'health.' Later, it will reveal that the cultivation of genetically improved species being is its essence. *Determinist* because it is power expressing itself in the predatory language of harvesting human flesh, and *transgenic* because it involves the capricious recombination of heretofore distinct species – firefly monkeys, jellyfish rabbits, headless organ 'donors' – transgenic determinism vivisects life into nothingness.

Physics may have been the privileged language of the atomic age, but biology is now the ruling vernacular of the era of completed technology. More than a technical language of the life sciences, biology expands now to become the dominant discourse of bio-politics: the framing language of capitalism, culture, politics, and media. No longer simply technological determinism, the order of the transgenic reflects something more subtle, equivocal, uncertain, and undecided. In genetic engineering, the order of the transgenic expresses itself in molecular freebase experiments where the genetic heritage of different species-types is capriciously recombined as primal models of the post-human. Streaming the genetic codes of animals, humans, and plants, the first inhabitants of post-human culture suddenly made their appearance: hybrid humans with the eyes of migratory birds capable of detecting invisible magnetic polarities in the sky; chip-enhanced bodies with slaved nervous systems centrally processed for better social control, always on standby for upgrading and virus protection; insect robots programmed with warrior genes, predatory instincts, and 'perverse' intelligence.

If molecular biology can adapt so quickly to the epistemological possibilities of the order of the transgenic, it may be because the spectre of transgenics originates less in the order of science than in culture. For quite some time, we have already lived within a deeply transgenic culture: in a culture of hybrid media images, streamed marketing practices, recombinant fashion, blended genders, a global skin of culture. *Transgenic imagery*: that is the 'bio-vision' of special-effects cinema or television ads playing with the digital referentials of space and speed to produce a visual universe of stretched vision, compressed skin, hybrid icons, performing cyborgs. *Transgenic capitalism*: that is the most recent fashion edition of Club Monaco, where, probably responding to the fear of ethnic difference in the air, the privileged colour is white accompanied by the ideological slogan: 'Finally have the courage to be the same.' Pilgrims of the future, we are already deeply habituated to the culture of the post-biological, to the language of post-human culture. We are only now in the infant stages of the order of the transgenic. The will to technology reveals itself under the sign of post-biologics.

Transgenic determinism codes the future of the twenty-first century, with Heidegger, Nietzsche, and Marx as simultaneously its fatal probes and most promising sites of resistance. *Heidegger, the philosopher of hyper-nihilism*, understands technology in the doubled language of techne and *poiesis*. His thought provides a bleak, comprehensive, and bitter account of the age of 'completed nihilism': its origins, development, and possible measures of recovery. *Nietzsche, the philosopher of cynical data*, thinks through to the other side of the will to power: not power in its accumulative stage, but power in its more dangerous contemporary historical phase as the 'will to not-will.' In Nietzsche's philosophy, the dynamic drive to the planetary technological future is revealed to be what Hannah Arendt once described as the 'process of annihilation' attendant upon the triumph of 'not-being.' Heidegger could accurately describe Nietzsche as the first philosopher of 'completed metaphysics' because in Nietzsche's thought the language of 'harvesting' and the reduction of human beings to 'passive standing-reserve' before the technological dynamo is expressed in its most vivid and chilling terms. *Marx, the philosopher of streamed capitalism*, awaits impatiently the historical self-realization of digital capitalism for the real radicality of his thought to be disclosed. Less a historically specific theory of the industrial phase of capitalism, Marx's *Capital* is a brilliant account of cybernetic capitalism: that moment when capital as a process of violent symbolic (digital) exchange refuses to separate itself from the cycle of circulation, migrating decisively beyond the obsolete models of pro-duction and consumption to the age of the digital commodity-form. In Marx's thought, streamed capitalism means that labour, knowledge, culture, power, and war are effectively networked together as different dimensions of a single global tendency towards the will to technology. Consequently, Heidegger's *hyper-nihilism*, Nietzsche's *cynical data*, and Marx's *streamed capitalism* as the new holy trinity of technology in the age of exhausted modernism.

Every mirror of culture has its flip side, and the unfolding destiny of technology is no different. If we are approaching an epochal rupture in the relationship between the body, culture, and society signalled by the order of the transgenic, then this story already contains a doubled possibility. A new order of values exists only to be reversed: that is its certain ontological fate. Consequently, if we now live through all the signs of a sustained drive towards transgenic determinism, we may also be the first witnesses of the *counter-blast*: a coming revolt by the order of the transgenic itself. In this case, does not the 'question of the transgenic' also pose a fatal challenge to the closed social constructions of modernity? *Transgenic bodies? Transgenic race? Transgenic gender? Transgenic art?* Might not the question of the transgenic objectively install a new order of floating identity, recombinant bodies, spliced gender, sequenced race that the binary divisions of the modern era have always functioned to erase? Will not the transgenic future also force a radical challenge to the question of the social construction of reality itself, to a series of antinomic, binary divisions that have always worked to exclude the radicality of the third term? There are dreams here – dreams of Nietzsche and Artaud and Bataille and Max Ernst and de Chirico and Duchamp and Burroughs and Acker, dreams of a third sex, a third body, a third identity, a third culture – dreams which until the advent of transgenic culture were always forced to the furthest margins of culture. This radical possibility, this *radical impossibility*, is also why an interpretation of recombinant genetics as the spearhead of the will to technology rubs together the reality of transgenic determinism against the hyper-reality of transgenic art. Once again, we find ourselves in the dwelling-place of the artist and the scientist – technology as craftsmanship and *poeisis*, capital accumulation and artistic imagination – opposing cultural terms which in their radical incompossibility mark out the present and future curvature of the will to technology as it migrates from the digital nerve to the genetic matrix.

4 Hyper-Heidegger: The Question of the Post-Human

Wireless in San Francisco

The sky is raining wireless in San Francisco.

Dispensing with the ordinary laws of nature, this city doesn't even wait for earthquakes to rebuild. Probably bored with the slow pace of the seismic faults, every two years it earthquakes itself: bodies, buildings, and businesses. Here, 'punctuated chaos' breaks out of evolutionary theory to become a strategy of life in the e-lane.

Just landed at SFO, Palm in hand, I check my email while waiting for my bags. Everyone else is doing the same. A corporate lawyer with that Palo Alto computer-burn look is saying something about trademarks and intellectual property rights into his cell phone. A guy in front of me flips through a book titled *The Power to Be Your Best*, while all the around the carousel the tech khaki look is definitely out, replaced by silver cell phones and black on black and tone on tone Armani suits. San Francisco is suddenly very LA, which probably means that digitality has reached that point of economic maturity where all the algorithms necessary for b-to-b commerce and struggles for market share have mutated the business cycle beyond drudge work to aesthetics. Art as the highest *expression* of technology, with wireless tech as itself the leading art form. At least this was true until the 'Slide.'

At the height of the dot.com boom, the drive into the city was straight propaganda for digital eyeball culture. Billboards everywhere streaked irises at a fast attention-span: 'Buy dot.com,' 'www.fatcity,' 'Real Time/ No Limits,' 'The Essential Element of the Internet,' 'Why Stream When You Can Burst,' 'It's the Animating Energy.' SF was a twenty-first-century technodrome in which every citizen was pressed into service as a stage prop in a gigantic sim/skin bubble that took over the whole (globalized) world. Local TV imprinted the language of tech biz on everything with stock quotes and minute-to-minute reports from the market front flashed over baseball games, accident scenes, political debates, movies, and soaps. Every passing car was a global info-scanner with one hand on the wheel and one ear tuned into the wireless future. Even the Embarcadero down by the just refurbished waterfront got into the act. As the *San Francisco Examiner* stated: 'Ride the S.F. Trolleys to L.A.'s Past.' And it was true. In a strange time reversal, the SF waterfront had been swelled up 1930s LA style: palm trees in the cold spring rain, cobblestone walkways, red brick faux nostalgia Pac Bell ball park, beautiful yellow and blue/green streetcars freshly imported from European dead tech graveyards to begin the new century with a tourist run from the Castro to Fisherman's Wharf. At the height of tech euphoria, San Francisco was the glistening capital of the e-commerce gold rush, pulsing with a doubled sense of manic buoyancy and a restless sense of anxiety that the virtual bubble of all the Lucents and 3Comms and Amazons of the tech business world was about to burst.

Culture of Boredom

After the collapse of tech euphoria, you find yourself one sunny afternoon sitting in a North Beach café just down the street from City Lights Bookstore, flipping through the pages of an art magazine and feeling the sense of abandonment in the air. The high-distortion bodies painted by Francis Bacon with their silent screams and twisted flesh warp right out of the photographs into the cultural psychology of the City. Max Ernst's dark, silent scene of *After the Fall, Silence* is a visual metaphor of the post-tech-hype times. Everywhere there's a fatigued sense that the party is over, small-time investors have gone home to the Midwest, the brilliant lights of tech hype are flipped off, and the only action lies in those eerie scenes of night-time reconstruction work, from the arc-lit steaming pit at the World Trade Center to Microsoft's installing the privatizing codes of intellectual property rights triumphant in X-Windows. The ecstasy of tech hype is replaced with a culture of boredom, cut with moments of intense anxiety and deep dread leavened by an entertainment complex that is definitely not down for long. I think of Heidegger: 'If, finally, metaphysics is the historical ground of the world history that is being determined by Europe and the West, then that world history is, in an entirely different sense, nihilistic.'[1]

In times of outer and inner crisis such as these, when our fundamental attunements to the times in which we live are suddenly put out of sync, when the symbolic universe has been wiped clean by the double terrorism of Islamic fundamentalism against the towers of Gotham and western state terrorism across the 'homeland' of globalization, it is sometimes useful to adopt the artistic practice of 'street-reading' initiated by the Montreal artist Francine Prévost. In her artistic imagination, the waste refuse of the streets of the global city, from Paris and Marseille to Louisiana and Montreal, these accidental remainders of process culture, has been assembled into a series of astonishing tableaus of the metaphysics of the twenty-first century. Images of torn advertisements, scraps of instant self-photos, handwritten addresses on envelopes, pages of a forgotten lover's diary, lines from scribbled songs never sung, broken dreams never spoken, framed portraits never remembered. Prévost's art is a contemporary version of Marcel Proust's *Remembrance of Things Past*, with the exception of this interesting twist. Rather than simply record that which is disappeared by contemporary culture, Prévost restages the remainders of city life to itself as an art of lost affections and silent whispers: a remembrance not of the past, but of what is to come. Which is perhaps why her act of street-reading is always doubled: at one moment a simple collection of that which has been tossed away and forgotten, *accidented* by culture, but also, and more importantly, visual memories that are recuperated in strings of aesthetically beautiful tableaus: torn shards of memory in which the compositional strategies of photography and the artifice of Japanese flower arranging are brought together to produce lines of memory traces, sometimes dozens of individual tableaus at a time, which when experienced directly and visually instantly attune your mind and emotions to the inner pulse of contemporary culture. Art as a street-reader of hyper-modernity.

With Prévost's 'street-reading' as my aesthetic strategy and with a search for a new 'attunement' to technology on my mind, I walk the streets of San Francisco, privileging the alleys around City Lights. If technology under the sign of business reproduced San Francisco as *Hollow City* in the 1990s, then the resistance to that speed machine must inevitably come from the young poets of the American heartland who, if the history of the poetic word is a guide, always break with suburban boredom to congregate again around the shrine of the American beats: Acker, Burroughs, Ferlinghetti, Ginsberg, Kerouac's *On the Road*. Not for purposes of mimesis, but to express in written language a new *Howl* fit for the age of viral terrorism. Like this fragment of a text which I scribbled and then tossed into a recycling bin at Columbia and Fine for another street-reader of the 'age of terror':

We stand abandoned. We live on empty. We get high on empty. We get low on empty. We are anxious. We are bored. The orifices of the body hip-hop to the ecstasy sounds of anger and desolation. Senses jam. Eyes edit/mix/mutate. The tongue swells with too much information, but sometimes too little to say. The tongue is a repeating machine stuck on empty. We are held out into the nothing. Aimlessness catches our drift. We are an experiment. We are the 'twilight of the idols.' We spew. We laugh. We are cells of gossip. We snack on images. We vomit. We watch the Web and download TV. We watch ourselves watching TV. We watch the screen watching us. We like the screen. It's our friend. It's our boredom. It's our crowd. It is our metaphysics. It is also the pleasure of forgetfulness. It is also how the nothing happens to us. But we love to be data-netted. We are the flesh of the new media. It is what is being done to us. But we love the nothing. We love the feeling of being held out into the void. We love what Heidegger warned us against: the absence of the 'essential oppressiveness' of dasein. We have reversed the optics of eyesight. Now eyes flip backward in the electronic skull to catch the inner scene of optic nerve. We are our distended eyes. We delight in watching ourselves being held out into the electronic void. We are driftwork spectators of our own emptying out. We are present at the catastrophe of the flesh void. But we are fascinated by catastrophe. It provides the illusion that something is ending. But *we* are only beginning. We have made simulacra of ourselves. The externalized eyeball of electronic culture captures us by the seduction of its speed. The digital tongue takes possession of our flesh. The digital nerve can program smiling faces and plastic teeth and targeted words because we are already the people of the void. We are the willing subjects of the digital universe. Digital survivors and cyber-code winners. We are the new economy. We're the new generation. We're new models of digital dasein. Digital being is being that is held out into the nothing. We have finally come home to nothing as the essence of our being. Heidegger might say that being is 'weakened in its ground,' but we find strength precisely in the weakening of our essence. We want to get rid of ourselves. We want to evacuate the flesh. We want to externalize eyesight and exteriorize the failing human sensorium. We are sick of being human. Technicity crushes. But we delight in the pleasure of being the first human beings to live with a crushed sensibility. We specialize in the sensual pleasures of cruelty. You will know us by our smirks. We are cynical. We are out to get what's ours. But we know it's just a game. Games kill, but that's the fun of it. Even conformity has its moments. We are circulating nodes – interfaces in the digital network. An interface has its private pleasures. Nodes are on speed. Nodes are on fire. Nodes are the wind and the liquid of the rivers and the tactile touch of the media spectrum. We are connected. We are distributed. We are circulated. We are wired. We are wireless. We are figured and reconfigured. Technicity is our subjectivity. That is our past, and our future.

Heidegger in Cyber-City

How to explain, then, this strange split in my thinking with one eye on the animating desolation of the wireless future and the other open to the lessons from the grave of Martin Heidegger, a thinker very much in political disrepute yet who, I am convinced, is the key philosopher of fully realized technological society, a theorist who provides both a fundamental metaphysics of virtual capital and a searing vision of the twisted pairs of desolation and freedom as technological destiny. Heidegger in San Francisco? Not past and future, but something much more primal: my body and mind and certainly my feelings as tech flesh moving fast into the learning nodes of mobile capital, fascinated by the radical global insurgency of it all, watching myself be captured as objectified flesh by a digital business machine that asks only that it servo-scan my iris, encrypt my signature, and record my electronic pathways in return for full participation in the games of net wealth. And yet, the deeper and faster I *am* wireless, some recalcitrant, maybe ancient, spirit of resistance keeps shutting down my vision machine, keeps pulling me away from an interior consent to my wireless desires. I'm wireless in San Francisco, bunkered down in a hotel room with all portals open to electronic ecstasy, mind-drifting those window images glorious of ships in the San Francisco Bay, yet I'm surrounded by book noise. Not just any disturbance of the written word, but a very specific set of texts that just might have something to tell me about my wireless ambitions: Heidegger's *The Fundamental Concepts of Metaphysics, The Question Concerning Technology, An Introduction to Metaphysics, Discourse on Thinking, Being and Time, The End of Philosophy,*[2] and, of course, his fateful retelling of the prophecies of *Nietzsche.*[3] It's philosophy as a radical experiment, with my body positioned in the most privileged space of global capital, and my consciousness given over to the deep time of metaphysics. Streaming cyberculture in the midst of the 'oblivion of the gods.'

The Art of Technology

It's all perfectly Heideggerian. After all, the real lesson of Heidegger is that metaphysics is the key to understanding the 'question of technology' as the sustaining vision of human destiny, a 'destining' which can neither be refused without abandoning the question of being itself nor be uncritically affirmed without opening the 'pathway' to human desolation.[4] If we can neither refuse technology since it is the essence of human identity nor stand quietly in the digital vortex without losing something essential to the deepest meaning of human existence, then we are in desperate need of a way of thinking the question of technology which mixes the language of destiny and contemplation. *A metaphysics of yes and no.*

In his *Discourse on Thinking*, Heidegger made an important double claim. First, he insisted that in a technological age typified by the 'flight from thinking' consequent upon the triumph of 'calculative thought' over meditation, what was of urgent ethical importance was the creation of a new 'fundamental attunement' to technicity.[5] An *art of technology*, therefore, which, developing in the midst of the most radical changes that technological society has to offer, would discover ways to 'release' the essence of technology towards its 'saving power.' In Heidegger's vision, calculative thinking represents only one tendency in technology as destiny. While the growth of calculative thinking dominates technological discourse, the essence of technology contains an equal, and until now hidden, countervailing tendency towards 'poeting technology,' releasing, that is, technology 'towards things.' In its essence, the meaning of technology is *split consciousness*: calculation versus meditation, objectification versus art, 'world' versus 'earth,' identity versus difference. Refusing an exclusive loyalty to either side of these polarities in the essence of technology, Heidegger's art of technology seeks to embrace both simultaneously. His thought is a radical experiment in intensifying the 'danger' in order to maximize the 'saving power.' He is the philosopher of the missing third term: eastern mysticism as the spectral absence of western rationalism. Consequently, Heidegger's project is this: developing a new *'comportment towards technology,'* as a way not only of understanding the underlying grammar of 'technology as destiny' but of negotiating the technological storm – 'poeting' technology with such eloquence and precision that out of the 'danger the saving power grows.'[6]

Second, Heidegger claimed as a continuing paradox of the human condition that to fully understand what is 'nearest' at hand, we often need to travel the furthest.[7] We understand the future best by exploring in depth the future of the past. Never a technophobe and certainly under no illusions that technology was anything other than the *essence* of being, he always insisted that the essence of what he liked to describe as the 'mystery of technology' was simply this – 'technology could not be understood technologically.'[8] While Heidegger was a thinker of his specific historical circumstances extending his meditations from the language of technology of the death camps of the 1930s to the atomic weapons of the post-war era and thereupon to an exploration of 'anxiety' and 'boredom' as the fundamental 'attunements' of contemporary techno-culture, so too 'overcoming' Heidegger, which is to say incorporating Heidegger's grisly vision of the 'world picture'[9] of technicity as an accomplished fact of contemporary society, means understanding what is nearest – wireless culture – in the language of what is furthest – the metaphysics of the 'mystery of technology.' Brushing against the immediacy of the wireless future intimates that the grounding conditions out of which wireless culture emerges as well as its future consequences for human culture can never be understood simply in terms of the narrow technological instruments of mobile communications, but only as the unfolding destiny of something much larger and as yet concealed in the very 'destining' of technology. Following Heidegger, what is needed is a metaphysics of the wireless future.

Now my wireless mind might have been possessed by the 'world picture' of 'bluetooth technology' and the speed integration of virtual capital, but my body in San Francisco emitted a different report. It was as if the flesh of my thinking had itself split into an irreconcilable doubling: a fully objectified mind living within the spectre of digital accelerants, moved by fascination and awe at the electronic creativity of it all, and yet feelings restless that bubbled to the surface of my distracted consciousness Heidegger's poeting of technology as 'destiny' in the chilling sense that today the gods have hidden themselves from us, abandoning us to walk the earth as desolation in the speed company of increasingly objectifiable beings. Or as Heidegger clarifies in the midst of the howl of the digital wind:

Yet it is not that the world is becoming entirely technical which is really uncanny. Far more uncanny is our being unprepared for this transformation, our inability to confront meditatively what is really dawning in this age.[10]

If Heidegger is correct, if, that is, it is not technicity in itself which is uncanny but the present degree of unpreparedness for *thinking* the meaning of technology as destiny that is uncanny, then it may also be that it is only by rubbing Heidegger against the technological dynamo – brushing the ancient language of metaphysics against digitality – that we can create a new ethics of technology. Against the prevailing political consensus of the time which continues, not without good reason, to refuse Heidegger, and certainly against the epistemological practice of situating Heidegger within the *history* of philosophy, I would like to make of Heidegger what might be called *thinking the uncanny*.[11] Thinking Heidegger, that is, as a way of 'dwelling' between the irreconcilabilities of the fully objectified beings of technological society and an as yet unappreciated, and certainly until now silenced, way of *being technology*. Uncanny thinking, therefore, as the first step in recovering an art of technology.

Uncanny Thinking

Martin Heidegger is the theorist par excellence of the digital future.

Probably because Heidegger's was a deeply embittered vision of the ruins of modernity to the extent that he wrote in a spirit of desolation about the 'gods having abandoned the earth,' retreating back into an impenetrable shroud of 'forgetfulness,' Heidegger was the one thinker who did not shrink from thinking through to its deepest depths the unfolding horizon of a culture of 'pure technicity.' While Heidegger began his writing with a deconstruction of conventional ontology in *Being and Time*, his lasting gift to the tradition of critical metaphysics was to perform in advance an intense, unforgiving, and unremitting deconstruction of his own life in *The Fundamental Concepts of Metaphysics: World, Finitude, Solitude*.[12] After the latter book, having nowhere to go other than to wander in the shadowland between a reflection on Being that in its retreat into forgetfulness was admittedly impossible to concretely realize and a future driven forward by the 'will to technicity,' Heidegger was the one thinker who literally deconstructed his own project to a point of self-nihilation. With nothing to save, no hope to dispense, and no critique that did not fall immediately into the dry ashes of cultural cynicism, Heidegger was fated to make of his own life of thought a simulacrum of the will to technology. More than Marx, who remained wedded to the biblical dream of proletarian redemption, and more than Nietzsche, who countered the nihilism of the 'will to power' with the possibilities of reclaimed human subjects as their own 'dancing stars,' Heidegger was the one thinker without hope in the dispensations of history.

Not broken by the vicissitudes of history, Heidegger was and is the contemporary historical moment. In his thought, the new century is already 'overcome' at the very moment of its inception. Not overcome in the sense of abandonment, but overcome to the extent that Heidegger summons up in his thinking the anxieties, fears, and methods of the will to technicity. A futurist without faith, a metaphysician without the will to believe, a philosopher opposed to reason, Heidegger is the perfect representative of the technological trajectory at the outer edge of its parabolic curvature through the dark spaces of the post-human future.

If it be objected that we should not read Heidegger because of his political complicity with German fascism, I would enter the dissent that Heidegger's momentary harmony, but harmony nonetheless, with the politics of fascism makes of him a representative guide to the next phase of fascism – virtual fascism. I would go further than liberal critics who fault Heidegger for taking advantage of the fascist upsurge in pre-War Germany to gain a university rectorship as well as to betray his philosophical mentor – Husserl – noting that in breaking with National Socialism, Heidegger did not refuse fascism on the grounds of an oppositional political ethics, but because its strictly *political determination* in the historically specific form of National Socialism in the Germany of the 1930s and 1940s was not a sufficiently 'pure' type to fully represent the metaphysical possibility that was the German 'folk.'[13] For Heidegger, National Socialists were not sufficiently self-conscious metaphysically, too trapped in the particularities of politics, to be capable finally of realizing the ontology of the fascist moment: delivering the metaphysical possibilities of (German) folk-community into concrete historical realization. To the tribal consciousness of fascism, Heidegger remained a metaphysician of Dasein. Ironically, his prescience concerning the fading away of second-order (National Socialist) fascism before the coming to be of first-order (virtual) fascism ultimately made of his thought a historical incommensurability: too metaphysically pure for the direct-action, 'hand-to-mouth' politics of German fascism; and yet too radically deconstructive of the claims of technological rationality to find its home in liberalism. *Homeless thought.*

An idealist in the tradition of German nationalism, Heidegger was fated to be the faithless thinker, ultimately disloyal to German fascism because it was not sufficiently metaphysical, yet unable to reconcile himself to western liberalism because it was, in his estimation, the political self-consciousness of technicity. For this reason, Heidegger ended the war digging ditches, having been ousted by German university authorities acting at the behest of state fascism as the University of Freiburg's 'most dispensable professor.' It is also for this reason that Heidegger in the post-war period was, except for a brief period before retirement, expelled from university teaching. Always a metaphysician, always in transition to the next historical stage of the 'will,' always in rebellion against the impurities of compromised philosophical vision, Heidegger was fully attuned to the restless stirrings of the will as it broke from its twin moorings in ethnic fundamentalism and industrial capitalism and began to project itself into world-history in the pure metaphysical form of the 'will to will.'[14] Beyond time and space, breaking through the skin of human culture, respecting no national borders, an 'overcoming' that first and foremost overcomes its own nostalgic yearnings for a final appearance in the theatre of representation, the *will to will*, what Heidegger would come to call the culture of 'pure technicity,' was the gleam on the post-human horizon, and Heidegger was its most faithful reporter. In Heidegger's writings, the main historical trends of the twenty-first century have their prophet and doomsayer.

Heidegger's mind lies between past and future.

Technology as a 'Danger' and a 'Saving Power'

If Heidegger could write so eloquently and think so mystically about that which in the present era is so unmentionable – Being – if Heidegger could say that Being 'comes into presence' in the mode of 'enframing,' the animating impulse of technology, if he could speak of Being as containing both a 'danger' and a 'saving power' and speak evocatively of the 'turning' so necessary to transform the danger into the saving power, perhaps that is because Heidegger's thought is itself a 'turning,' a 'lightning-flash' which illuminates human beings to themselves, and which does so not by surrendering to calculative thinking or by retreating to spurious forms of idealism, but by looking deeply and meditatively into the danger of technology, by 'thinking' technology to its roots in metaphysics.

Hyper-Heidegger, then, a thinker who makes of himself both a 'danger' and a 'saving-power,' who makes of the effort of reading Heidegger a form both of 'unconcealedness' and 'openness.' If Heidegger could dismiss as illusory thinking the pretension that 'man has mastery of technology,' claiming instead the opposite that *human beings are set in place as a condition of possibility for the development of technology*,[15] if Heidegger could only speak of the human essence in terms of its deep entanglement with the question of technology, that is because Heidegger's thought is the 'clearing' that he thought he was only prophesying. To read Heidegger is not so much a matter of meditating on the 'question of technology,' but the much more dangerous possibil-ity of becoming entangled with the *question of Heidegger*. Not Heidegger as a historically proximate philosopher with a certain biography as a determinately local German thinker projecting the 'pathways' of the Black Forest onto the 'world picture,' but Heidegger as that 'glancing' taking us immediately into the dangerous mysteries, not of being, but of *hyper-being*, into the impossible metaphysical claims of a form of being that only exists in the language of fatal oppositions: calculation versus meditation, world versus earth, ordering versus revealing, technicity versus art. Refusing the safety of a strictly monistic determi-nation of the question of being, Heidegger was always a hyper-metaphysician, making of being an enigmatic sign, a crossing-over, a 'solitude' between the identity of 'world' and the difference of 'earth.' For him, *incommensurability* is the essence of technology, and hyper-being the song-line of the deeply conflicting impulses that ani-mate technological destining.

The question of Heidegger necessarily speaks to the human essence. If Heidegger is correct, the discourse, first of capitalism, then of capitalism in its hyper-phase as virtuality, is the story of the presencing of *hyper-being*, with ourselves as both its active participants and necessary condi-tions. This is not a story of fatalism or catastrophe – far from it, since Heidegger claims human subjectivity is not only a 'historiographical' representation of technological consciousness – but also the story of 'destining,' of learning a certain 'comportment towards technology' that draws the saving-power out of the danger of technology. In the strange labyrinth of history, could it be that the question of Heidegger is also a 'turning,' a way of looking deeply into the danger as the first tentative steps towards the presencing of another destiny of technology? Heidegger went to his death with the constant admonition that we are 'uninterpreted signs.'[16] Could it be that interpreting Heidegger is the necessary encryption of the codes of technology, that until now neglected interpre-tation of the 'uninterpreted sign' that is digital being? But, if that is so, if

Heidegger is the necessary interpretation of technological destining, then wouldn't that also make Heidegger's thought a form of 'valuing,' a will to power projecting itself across the world picture in the language of thought? Wouldn't Heidegger's destiny, then, be an *artistic* one: simultaneously fully implicated in the question of technology while different from it, an artist of the 'yes and no'?

Out of place in his time, a thinker sensitive to the loss of the autochthonous in the culture of technicity, Heidegger transformed the language of *'rootlessness'*[17] into a central premise of the strife in modern subjectivity. For him, the challenge and impossibility of the modern technical project was its starting-point in *'being held out into the nothing.'* Camus's *absurd*. The gods have retreated into the shadows. The meaning of technicity lies close at hand, yet remains concealed in the shroud of calculative forgetfulness. No certain past, no actual present, only a future-time split open by the animating energy of the will to technology: cultural 'rootlessness' as the central feature of modern technical being. Indeed, if contemporary subjectivity can move with such volatility between the 'malice of rage' and the solace of healing, then this would only indicate that strife is the modern language of rootlessness. This, then, is the modern fate: 'being held out into the nothing' with no clear way of returning to oneself as an abode or dwelling in proximity to the ancient language of the 'holy.'[18] And yet if we cannot think of the self as an abode or dwelling, then what remains is only the desolation of homelessness and its certain result – the 'malice of rage.' For Heidegger, as earlier for Nietzsche, who in *On the Genealogy of Morals* spoke evocatively of modern being rubbing itself raw on the bars of 'civilized' culture, the 'malice of rage' is the true malignancy of technological culture. That this malignancy can sometimes be distracted, even to the point of forgetfulness, in the form of technological exteriorizations of the human sensorium and, at other times, temporarily appeased in the sacrificial language of ethnic scapegoating, does not dispense with the sense of strife central to technical being. If we are an 'uninterpreted sign' projected into the future and concealed from the past, then the malignancy at the core of technicity might itself, if intensified by thinking, be compelled to reveal its essence. Which is, of course, the value of contemplating Heidegger: a thinker so proximate to the contemporary technical condition that his thought is itself a field of strife, motivated from within by a malice of rage directed against his own expulsion from the polity of conventional political opinion, and yet a thinker who, in the bitterness of this exile and undoubtedly against his own preference for the rootedness of the 'German folk,' became a vehicle by which the forgotten language of metaphysics – the homeward-bound language of the pre-Socratics – speaks again to beings held out into the nothing.

In contemplating Heidegger, we also return to ourselves as 'uninterpreted signs.' His writing is the future of the past.

Philosophy of Technology

> All that is merely technological never arrives at the essence of technology. It cannot even recognize its outer precincts.[19]

Make no mistake. Heidegger does not 'think' technology within its own terms. Quite the contrary. Repeatedly he insists that technology cannot be understood technologically because, in opening ourselves up to the question of technology, we are suddenly brought into the presence of that which has always been allowed to lie silent because it is the overshadowing *default condition* of our technical existence. Heidegger is relentless in making visible that which would prefer to remain in the shadows as the regulating architecture of contemporary existence. For example, Heidegger notes that, today, we can only think technology from the midst of the howling centre of the technological vortex, that while we can note that the dominant tendency of technology is towards the 'objectification of earth' and the 'objectification of (technical) consciousness,'[20] we can never be confident that, in thinking the consequences of technologies of objectification, our thought itself has not already been set in place as a necessary '*turning*' of the technological spiral. And while Heidegger will note that the key ethical consequence of the relentless objectification of earth and sky and water and flesh is '*injurious neglect of the thing*,'[21] he always makes the parallel claim that thought itself always has about it a form of neglect, that thought, however critical, always conceals and unconceals, that 'injurious neglect of the thing' in the mode of order of willing and doing may also have about it the doubled language of human destining. Thinking Heidegger from the virtual present, from the perspective of the 'shadow cast ahead by the advent of this turning,'[22] that he could only intimate, who cannot be fully ambivalent on the ultimate meaning of technology as 'injurious neglect of the thing.' *Who, that is, cannot brush thought against that doubled possibility of injurious neglect, that such injurious neglect may be, in equal parts, a brutalizing consequence of the dynamic language of (technical) ordering and willing and the deepest seduction of technol-*

ogy? In this case, if the price to be paid for the unfolding of (our) technological destiny is 'injurious neglect of the thing' to the point of gutting human subjectivity of its silences, its most essential elements of individual reflection, of thoughtfulness, then is it not now manifest that such injurious neglect *of oneself* is the deepest fascination and most charismatic promotional feature of virtual capitalism? *The virtual self, therefore, as a wireless game with accelerated technical consciousness moving at the speed of injurious neglect.*

Consequently, Heidegger's specific contribution to understanding technology consists of a unique, evocative, and comprehensive description of technological experience as a single human process originating in the metaphysics of 'enframing,' driven forward by the animating energy of the 'will to will,' resulting in a culture of 'profound boredom,'[23] and possessing art as its possible 'turning.' Folding together future and past, Heidegger's theory of technology assumes the form of a general theory of civilization which, beginning with the basic assumption that technology cannot be understood solely in the language of the technological, traces the genealogy of 'planetary technicity' to its ancient roots in a *way of being* that, expanding from its origins in the mythic legacy of the west, comes to represent human destiny. As human destiny, technology can neither be refused nor simply affirmed because of its inextricably ambivalent nature. Left unquestioned, technological experience reduces life to a 'standing-reserve,' in the 'unconditional service' of the will to technique. And yet if the 'question of technology' cannot be asked without a fundamental inquiry into the mythic roots of technology as destiny, then it must also be said that the (hyper)reality of technology cannot be denied without a fateful loss of that which is fundamental to humans qua humans. For better and for worse, in boredom as well as in anxiety, the question of technology as destiny means that it is only by intensifying technology, by 'thinking' technique to its roots in ancient mythology and, thereupon, to its future in the expanding empire of 'planetary technicity,' that we can hope to elucidate the dangers and possibilities of *being human* in the dawning age of the post-human. Heidegger's 'question of technology' is also a way of coming home to the neglected question of the meaning of life in the technodrome.

The Politics of the 'Standing-Reserve'

Heidegger's famous essay 'The Question Concerning Technology' can only be read now in terms of philosophical anthropology. Against its own intentions, which were focused on stripping away history from the question of technology and, thereupon, grounding the question of technology in the language of its founding metaphysics, this essay has in the years since its authorship been reclaimed by the riddle of history. Reclaimed, that is, not in the sense of obsolescence – a theory of technology now superseded by accelerating developments in the present age of wireless and biogenetic invention – but reclaimed in the deeply anthropological sense that Heidegger's analysis of the question of technology is an uncannily accurate diagnosis of the present human situation.

Although he wrote from the perspective of a mid-twentieth-century historical period bracketed by the rise to dominance of mechanical technologies of extraction and the overpowering presence of atomic weapons, Heidegger's view of technology, while focused on mechanical culture, only finds real theoretical and ethical purchase with the advent of electronic and, thereupon, digital culture. In a way that foreshadows contemporary theories of technology, from Virilio's vision of cybernetic technology as a 'war machine' operating in the language of the control of 'eyeball culture' and McLuhan's grim vision of the 'externalization' of the central nervous system in electronic culture to Baudrillard's theorization of the mass simulation of human desire, Heidegger does that which is most difficult. Almost as a precession of his own theory, his analysis *presences* technology, drawing out the animating impulses of techno-culture in such a way as to compel the 'world picture' of technology to fully reveal itself. Refusing to think technology separately from the question of human destiny, Heidegger's thought always hovers around two conflicting impulses in the technological world picture: first, the tendency towards 'enframing' by which the dominating impulse of contemporary technology pirates the human sensorium on behalf of a globally hegemonic technical apparatus; and, second, the tendency towards 'poeisis' by which an art of technology, variously expressed in language, poetry, the visual arts, speed writing, an aesthetics of digital dirt, and new media art, could draw out of the world picture of technology as destining a different future for techne, a future in which technology once again has something to say, to 'unconceal,' about the relationship between technology and alethia (truth).[24]

Indeed, what is so inspiring about Heidegger's doubled vision of technology is its uniqueness in simultaneously running parallel to the cutting edge of new digital technologies and doing so in such a way as to plunge the 'question concerning technology' back into its classical origins as a essential expression of being itself. While other theorists have 'thought' technology within and against the modernist and now postmodern epistemes, Heidegger's special gift to those intent on deciphering the question of technology is a dramatic double refusal: refusing, at first, to think technology within strictly contemporary terms by insisting that the language of technique is derivative from another, more hidden, 'presencing' of being that hides itself in the shadows of thought; and refusing to think technology as technology, insisting that technology is at its inception never strictly technological but metaphysical.

Consequently, the curiosity: Heidegger's 'The Question Concerning Technology' makes of the dynamic drive to planetary technicity a probe for unconcealing a more fundamental 'mode of being,' a mode of being which, until now, may have purposively retreated into the shadows in the spectral form of 'oblivion of being,' but which under the artistic 'revealing' that is Heidegger's method is finally forced to confess its ancient secrets. In Heidegger's vision of technology, we are always standing midway between the unfolding future of the drive to technological domination and the revelation of the classical genealogy of the question of technology. Both genealogist and futurist – artist and craftsman – Heidegger probes the 'world picture of technology' in a way that is always enunciated in the doubled language of that which he seeks to expose – the twin words of provocation and revelation, 'challenging-forth' and 'poeisis.' He is instructive to meditate upon not simply for his dramatic political and cultural conclusions concerning the destiny of technology, but, more decisively, for the deep method of his thought. Always equal to the object of his writing – planetary technicity – Heidegger not only claimed that technological experience was, above all, a *method*, but in his own writing paralleled the world picture of technology as method by making of his own thought a method of technological revelation. In meditating upon Heidegger, we are suddenly brought (technically) close to that which is (metaphysically) distant. His mind splits the atom of technology. His thought sequences the DNA of the question of technology.

In Heidegger's thought, the twin elements composing the atom of technology in its classical origins and which, until now, have wandered the 'desolation of the earth' separate and at war, these twin elements of provocation and poeting, calculation and meditation, space and time, are finally reunited in a new experimental moment of fusion. The Heideggerian method solves the riddle that it sought only to reveal and, in doing so, provides an *ethics of technology*, an ethics that has something fundamental to say about the unfolding future of planetary technicity because the Heideggerian project *is* technology. Beyond the specific historical details populating each of Heidegger's writings on technology, from the atomic weaponry of 'The Question Concerning Technology' and the theoretical physics of 'What Is Metaphysics?' to the biogenetics of *The Fundamental Concepts of Metaphysics*, Heidegger brings to the project of thinking technology a mode of expression simultaneously ancient and post-human, equally at home in the question of being and not-being. And if, at the end of his life, Heidegger abandons the comfortable illusions of existentialism that are the condition of possibility of *Being and Time*, that is only because, faithful to the method of 'challenging-forth into the ordering of the standing-reserve'[25] that is the hallmark of the technological surgery upon the human condition, Heidegger does not, in the end, spare his own thought from the bitter lessons of his diagnosis. This is one thinker with the courage to make of his own theory of technology a model of technicity with such intensity and determination that his thought challenges technology to the death. Challenges, that is, the world- picture of technology to circle back on itself, to engage the conflicting impulses towards 'harvesting' and 'poiesis' in their most primary expression of being in Heidegger's 'way of thinking.' Without exaggeration, the *alethia* – the truth – of Heidegger is, at once, the *alethia* of technology. Resolving the limits and creative intensities of Heidegger's vision of technology is much more than another perspective external to technology. To *think* Heidegger is also to presence the interior limits of a mode of (technical) being that seduces by its radical impossibility: revelation without actualization, calculation by abandoning justice to the oblivion of being. The question of Heidegger is proximate to understanding the twenty-first century.

Vampire Metaphysics

Heidegger presents us with a metaphysics of political economy. Beginning with the assumption that political economy is not understandable solely in its own language, Heidegger describes a relentless politics of economic and psychological appropriation by which the world-picture is reduced to a machinery of harvesting. Everything is there: the reduction of human experience to a 'standing-reserve'; the mobilization of human consciousness into a support-system for technicity; the muta-

tion of human flesh into the skin of the technodrome; the coming alive
of technicity as a disembodied cybernetic organism, part-flesh/part-
machine; the 'harvesting' of human vision as a cybernetic steering-
system for the new economy. Here, the global political economy is
rapidly transformed into an 'energy source' for the coming to be of the
fully realized technological future, with everything in a permanent
'waiting' mode, on stand-by ready to be parasited by the demands of
technicity. The French theorist Paul Virilio might have described the
politics of 'electrooptics' as a form of dromology, but Heidegger went
further. For him, contemporary political economy is the exterminatory
metaphysics of harvesting the 'standing-reserve' (humans, animals,
nature) of its living energy, and then abandoning it as yet another
empty node in a random cycle of economic circulation. In Heidegger's
sense, contemporary political economy is understandable only in the
language of vampire metaphysics.

> Everywhere, everything is ordered to stand by, to be immediately on
> hand, indeed to stand there just so that it may be on call for a further
> ordering. Whatever is ordered about in this way has its own standing. We
> call it the standing-reserve.[26]

All inescapable and all-too-(post)human, because of Heidegger's still
undigested insight that technology is human destiny, that the question
of technology cannot be understood technologically, only metaphysically. The spectre of the vampire haunts the language of the 'standing-
reserve.'
   The revealing that rules through modern technology has the character of a setting-upon, in the sense of a challenging-forth. That challenging happens in that the energy concealed in nature is unlocked, what is
unlocked is transformed, what is transformed is stored up, what is
stored up is, in turn, distributed, and what is distributed is switched
about ever anew. Unlocking, transforming, storing, distributing, and
switching about are ways of revealing. But the revealing never comes to
an end. Neither does it run off into the indeterminate. The revealing
reveals to itself its own manifoldly interlocking paths, through regulating their course. Regulating and securing even become the chief characteristic of the challenging revealing.[27]
   This is the essential point. If, for Heidegger, the essence of technology
does not lie in technology, this would only mean that in *being technological*, in identifying our fate so implacably and urgently with the will to

technology, we are swept up in a broader destiny – a 'destining' not of exclusively human origin, certainly not under human control, and most definitely not brought out of concealment into the clarity of thinking. Thus, for example, while Marx would reduce technological experience to the cycle of capitalist production for its explanation and while present-day technotopians would confine an understanding of virtuality to technical advances in chip speed and networked solutions and connectivity, Heidegger's claim is the more audacious, and yet more potentially disturbing, and ennobling, of technology.

Switching Flesh

In a way that is deeply mystical, Heidegger says of the essence of technology that it is directly related to the question of being. In an age that prides itself on its spirit of economic deconstruction, that advances by vivisecting the question of origins into oblivion, Heidegger is the one thinker who remains entranced by the stubborn fact of being. Not being in a narrow religious sense or as a repository of the past, but being as something more pervasive, and for that reason more enduring in human experience. Always close at hand yet concealing itself in the shadows, the question of being can be approached only through its secondary effects, in terms, that is, of how being manifests itself. To speak the language of destiny does not imply being dragged along by the blindness of fate or being without choice in the unfolding events of human history, but it does mean being open to the possibility that technology is a secondary-effect, a *condition of possibility*, that is set in place as a way of disclosing the 'mystery of being.'[28]

Of course, all such language pertaining to the mystery of being is forbidden in the rush to claim membership in the empire of the virtual will, but, perhaps, if we cannot think technology to its roots in the mystery of being then it may be that we will have to settle ourselves for a secondary life as we are reduced to unselfconscious conditions of possibility for the operation of the technical apparatus. In Heidegger's terms, thinking technology in its essence as a way of revealing the mystery of being does not imply a 'blind compulsion' to follow the technological imperative or a futile spirit of 'rebellion,' but constitutes an opening to freedom. In this case, listening to the intimations of experience released by the opening to technology may take us furthest towards that which has been always close at hand, allowing us to travel to the past of the mystery of being in order to understand the future of the mysteries of life. Freedom, in this case, would have everything to do with working through the conflicting impulses in the technological dynamo. Since we cannot retreat from technology because, as a matter of being itself, it is our destiny and since we cannot quietly acquiesce in the imperatives of technology as a form of aggressive appropriation that would reduce us to 'standing-reserve,' then the question of freedom implies working through the mystery of technology.

Between technology as vampire metaphysics and an 'opening to freedom': that is the essence of the mystery of technology. Heidegger is insightful when he speaks of the origins of technology in what he describes as 'enframing.' Not enframing in the strict etymological sense of a 'building' or a 'skeleton,' but enframing as an ancient impulse by which being addresses us, when, as a matter of the human essence, albeit in its strictly western determination, we are 'challenged-forth' through technology 'to order the real into the standing-reserve.'[29] Technology as appropriation is that which is most intimate, and being so intimate, essential to our being. For better and for worse, human destiny is deeply entwined with its metaphysical origins in being 'challenged-forth' to reveal the real as 'standing-reserve,' to 'set upon' human flesh and nature in such a way as to appropriate all living energy on behalf of a cycle *of challenging, transforming, storing up, transforming, distributing, and switching about.* Understood as revelation, what technology as appropriation first discloses is an almost irresistible tendency towards the virtualization of human experience. Not only the virtualization of nature by an economic cycle that moves in a circular flow from challenging to switching – from expropriation of energy to indeterminate connectivity – but the virtualization of human flesh itself as the body, too, is set-upon by technologies of exteriorization, harvested of its energies through a random technical cycle that Heidegger calls '*challenging-forth,*' but which, for example, we now know most familiarly as the twin calculus of global capitalism and bio-engineering. Indeed, that acceding to the challenge of disclosing our essence as 'standing-reserve' is so essential, and consequently so comfortable, in the unfolding entwinement of human and technical destiny is revealed daily by the spectacular ease with which consumer flesh is prodded by machine culture into an empty node on the network. *Brand culture.* While Heidegger, writing at the advent of the flight of the gods from the earth, may have had confidence in the restorative dualism of meditation versus calculation, the panic buoyancy by which human being networks itself, is bio-engineered, camcorded, surveillanced, and netted suggests that in this time after-Heidegger, human society may have now become forgetful, if not oblivious, to its standing role as a 'condition of possibility' in movements of the virtual will. But if we are forgetful, if we willingly assent to the expropriation of our (electronic) attention by the machineries of (digital) harvesting, this is not to deny the destiny of technology as something essential to our being and, for that reason, a matter of neither passive assent nor futile rebellion. With and against the mystery of technology, we need also to invoke the mystery of art. In the time of vampire metaphysics, it is only by invoking what is hidden by the glare of technological appropriation that we can, hopefully, discover another pathway into the 'essence of technology.'

Perhaps this is why Heidegger ends 'The Question Concerning Technology' on a redemptive note. Mindful that for the Greeks technology was always equal parts techne (craftsmanship) and poeisis (art), Heidegger says that the destiny of technology is ambiguous. Ambiguous, that is, because technology as a manifestation of something more primal in being does not consist solely of enframing, but also of the revealing of the possible 'truths' of technology. In this, he is explicit:

> The irresistibility of ordering and the restraint of the saving power draw past each other like the paths of two stars in the course of the heavens. But precisely this, their passing by, is the hidden side of their nearness.[30]

The coming to presence of technology threatens revealing, 'threatens it with the possibility that all revealing will be consumed in ordering and that everything will present itself only in the unconcealedness of standing-reserve.'[31] But, in these circumstances, this danger can never be confronted directly or exiled from the human destiny. It can only be confronted by reflection.

> There was a time when it was not technology alone that bore the name techne. Once that revealing that brings forth truth into the splendor of radiant appearing also was called techne.

> Once there was a time when the bringing-forth of the true into the beautiful was called techne. And the poeisis of the fine arts also was called techne.[32]

Refusing nostalgia for the Greeks of antiquity and yet not assenting to the harvesting of art by the will to technology, Heidegger provides this important lifeline out of the technological whirlpool. Since the essence of technology is never strictly technological, the special contribution of art, today, is to begin such a fundamental reflection upon technology and 'decisive confrontation with it.' Art in its doubled appearance as reflecting upon and confronting technology approaches once again its classic promise to meet the mystery of technology with the mystery of the artistic imagination. Simultaneously 'akin to the essence of technology' and yet 'fundamentally different from it,' art is the other, hidden destiny of the will to planetary technicity.

The 'Will to Will'

> ... the correctness of the will to will is the unconditional and complete guarantee of itself. What is in accordance with the will is correct and in order. In this self-guaranteeing of the will to will the primal truth of being is lost.[33]

In the same way that Heidegger said that 'man is a transition,' a 'going between' the past and the future, so too all of Heidegger's thought might be considered a projection of the future of 'planetary technicity.' Here, Heidegger's philosophy of technology is itself a transition between the modern century conceived as a 'going away' and a future not yet disclosed. While Heidegger's fundamental work, *Being and Time*, might appeal to dasein (being) as temporality, the legacy of Heidegger is to have prophesied the disappearance of technical being into the pure space (of digitality). While a modern era dominated by the problem of temporality might come to be known as an 'age of anxiety,' a postmodern epoch constituted by the sovereignty of (electronic) space has profound boredom as its 'fundamental attunement.' For Heidegger, the contemporary period overcomes the oscillation between the past as 'going away' and the future as unrealized possibility in favour of something radically different: temporality as 'standing still.' The transition is finally finished; we live in the non-time and non-space of 'completed metaphysics.'[34] Completed, that is, because the scaffolding of external historical drives, from capitalism to religion, drops away, leaving the edifice of a fully realized technical society which has 'aimlessness' as its aim and 'using up' as its method.

In *The End of Philosophy*, Heidegger insists that ours, this age of hyperrationality that has supposedly freed itself from the constraints of mythology, is, in fact, the most intensively mythic of all ages. Mythic, that

is, not in the sense of either a nostalgic reprise of the history of mythology or the creation of a new pantheon of digital myths, but mythological in the most powerful sense that ours is the age of 'technology as
completed metaphysics.' In the technical epoch, the language of metaphysics can pass into oblivion because metaphysics is now everywhere:
simultaneously the 'guarantee of stability (truth)' of technological society as well as its 'exaggerating drives (art).'[35] The twenty-first century
will be the metaphysical century without limits.

Beyond Nietzsche's reflections on the will to power as the animating
energy of history and refusing, by default, Marx's analysis of capitalist
accumulation as the essence of historical materialism, Heidegger proposed that contemporary society is a vast materialization of a fundamental metaphysical force: the 'will to will.' Here, the will is stripped of
its motivating referents. Neither Nietzsche's will to power nor Marx's
will to accumulation, the will to will is the *completed* will. For Heidegger,
we live in the age of the *virtual will* in which the 'will rigidifies everything in a lack of will.'[36] Beyond the language of representation, the will
to will is virtual because it functions by continuously folding back on
itself in an indefinite recursive spiral. Again, neither past nor future,
but temporality as 'standing still.' Abandoning the language of representation as secondary effects,

> [t]he will to will forces the cancellation and arrangement of everything for
> itself as the basic forms of appearance only, however, for the uncondition-
> ally protractible guarantee of itself.[37]

What dominates, then, is the *virtualization* of the will – this new historical phase we are now entering in which everything will be calculated
and arranged as 'guarantees' of the triumph of the will to will. No
longer the *will to* anything, but now only the will ordering every dimension of life in order to sustain its own existence. This is why, for
Heidegger, ours is the age of 'completed metaphysics.'

> The past means here: to perish and enter what has been. In that metaphys-
> ics perishes, it has been. The past does not exclude, but rather includes the
> fact that metaphysics is now for the first time beginning its unconditional
> rule in beings themselves, of what is real and of objects ...[38]

What does it mean for metaphysics to begin its 'unconditional rule in
beings themselves'? This is the decisive question. In Heidegger's vi-

sion, a fully realized technological culture is not understandable except in relation to the question of metaphysics for the reason that *technology is metaphysics*. Technology is the 'guarantee of stability (truth)' of the will to will which together with the 'exaggerating drives (art)' is the dynamic instrument by which life is 'ordered' on behalf of completed metaphysics. Consequently, the age of 'planetary technicity' needs to be reinterpreted as an age not of hyper-realism, but of the deep infiltration of metaphysics into consciousness, flesh, drives, and desires. Digital being is fully realized metaphysical being, just as much as the hyper-rational consciousness necessary to digital existence is heightened metaphysical consciousness. Nothing escapes the will to will. It is the standard operating code driving forward digital being, and on behalf of which the virtualization of experience proceeds. Digital existence can be so dynamic, the 'new economy' can triumph over old economies, connectivity can hyper-inflate the speed of communication, and a new class of planetary technical intelligentsia can attain economic and, then, political power because all of these are *realized metaphysics*: instruments which are 'correct' because they are acting in accordance with the 'will to will as the unconditional and complete guarantee of itself.' In the technical epoch, the language of metaphysics can pass into oblivion because metaphysics is now everywhere. Thus, for example, the virtual class is driven forward by the will to will of which it is both subject and object: *subject* because the virtual class benefits directly from the unconditional attainment of the will to will; and *object* because the virtual class is manufactured as the raw resource necessary for the transition to the will to will. What Heidegger said of human beings as always standing restless between past and future could be applied even more intensely to the virtual class. It is the class of 'ubiquitous uncertainty': a transition, a dynamic (business) method which 'has become the very objectivity of its objects.' As the historical realization of the will to will, the virtual class is the metaphysical class par excellence. Its time, the time of speed and frenzy, is, in reality, the time of the 'standing now.' Its space, the historical space of the spearhead of technicity, is, in actuality, the space of the 'unhistorical.' Willed by the will to will, the virtual class is the 'accomplishment of striving,' and, as such, its dominance suggests that metaphysics has already entered its transition to another possibility.

Nietzsche warned in *The Will to Power* of a future age in which everything would be transformed into a 'mere condition of possibility.' A post-human history in which the will would drop its referents (power, accumulation, desire, consciousness), and suddenly perform a metaphysical gamble, revolving around itself as its own 'unconditional' and 'complete guarantee.' This, then, is the real meaning of the new economy. Not virtual capital understood simply as an extension of the order of logic associated with traditional (finance or industrial) capital, but a form of capitalism that can so quickly achieve its ascendant historical apogee because it is only derivatively about capitalism, and essentially about metaphysics. Capital, too, is placed in the service of the virtual will as a 'mere condition of possibility.' Virtual capital is the fully realized consciousness of metaphysics. It belongs to the will since the will to will itself is 'the highest and unconditional consciousness of the calculating, self-guaranteeing of calculation.' Here, virtual capital is not so much in the service of the sacred but of the 'correct.' As Heidegger reflects, 'correctness' in terms of the alignment of particular human wills with the will to will is the key political value of epistemology in the technical epoch.

> What is in accordance with the will is correct and in order. In this self-guaranteeing of the will to will the primal being of truth is lost. This correctness of the will to will is absolutely untrue. The correctness of the untrue has its own irresistibility in the scope of the will to will ... What is correct masters what is true and sets truth aside. The will to will unconditional guaranteeing first causes ubiquitous uncertainty to appear.[39]

'Ubiquitous uncertainty'? In the same way that Marx could write of the relentless movement of the commodity-form, Heidegger says of technology as completed metaphysics that the creation of radical uncertainty is its standard operating procedure:

> ... ubiquitous, continual, unconditional investigation of means, grounds, hindrances, the miscalculating exchange and plotting of goals, deceptiveness and maneuvers, the inquisitorial, as a consequence of which the will to will is distrustful towards itself, and thinks of nothing else than the guaranteeing of itself as power itself.[40]

Thinking Nihilism Nihilistically

In this vision, everything becomes 'raw material' for the will to will: *consumption machines* (being 'used up in consumption'); *code machines* (the 'self-release of being into machinations'); *speed machines* ('let[ting] his will be unconditionally equated with this process, and thus at the same time, becom[ing] the 'object of this abandonment'); *virtual machines* ('technology is the organization of a lack'). This, therefore, is the real significance of the cycle of exchange prefigured by virtual capital. As raw material, the virtual class is driven forward by the will to will of which it is both subject and object. It's the very same for the most privileged participants in the new economy. Not the origin of this historical era, they are 'willed by the will to will' as its most important raw material.

As for the will to will itself, *it is a* virtual will *because it has no necessary goal.* It is trans-historical and unworldly: *trans-historical* because the will to will injects aimlessness into the circulatory systems of the social; *unworldly* because the will to will 'absolutely denies every goal and only admits goals as a means to outwit itself willfully and to make room for this game.'[41] In the game of virtuality, the will to will doesn't wish to appear as an 'anarchy of catastrophe,' so it must continually legitimate itself: the will to will invents here the talk about 'mission.' Understood metaphysically, the 'mission statements' of businesses in the new economy are always second-order alibis, providing the illusion of a necessary purpose to what is essentially aimless and directionless. As Heidegger states:

> ... this struggle (for power) is in the service of power and is willed by it. Power has overpowered these struggles in advance ... This struggle is planetary and undecidable because it has nothing to decide ... through its own force it is driven out into what is without destiny ... into the abandonment of being.[42]

But if the game of the virtual will is 'planetary' and 'undecidable' because it has 'nothing to decide,' then this would mean that the contemporary period is novel to the extent that human beings today are compelled by force of (metaphysical) circumstance to live out their lives within the unfolding destiny of the 'abandonment of being.' While once we might have thought nihilism from the outside of culture, today nihilism in the sense of an abiding 'lack,' an 'emptiness of being,' is the deepest internal drive of society. The virtual will, this emblematic sign of technology as 'completed metaphysics,' has immediate moral consequences. Heidegger asks: 'What kind of humanity is capable of the unconditional completion of nihilism?'[43] What kind of humanity will emerge from the 'consumption of beings for the manufacturing of technology'? While beings may release themselves into 'machinations' and, consequently, place themselves in the 'unconditional service' of the virtual will, what of the more urgent question: what does it mean to be suspended in 'complete emptiness'? to live within the hovering shadow of an 'emptiness that has to be filled up'? to be in the unconditional servitude of 'technology as the organization of a lack'? What does it really mean to live out as human destiny the *undecidability* of the will to will, to acknowledge with Heidegger that *aimlessness*, 'the essential aimlessness of the unconditional will to will, is the completion of the being of will incipient in Kant's concept of practical reason as pure will'?[44] Indeed, is it only an anticipatory sign of the human inability, or perhaps unwillingness, to think through the consequences of the abandonment of being prefigured by the virtualization of the will that 'thoughtlessness' and the celebration of 'unreflective experience' are trademarks of the new economy? *Today nihilism can only be thought nihilistically* because all thought is touched by the emptiness of (technological) being, either as its promotional content or mode of concealment.

And yet, if we are fated to experience the full meaning of the destiny of technology as completed metaphysics, if we are destined to be fully proximate to the implications of the abandonment of being as the necessary condition of possibility of the new economy, then this would also indicate that a counter-movement in this transitional period is already in motion. If technology is human destiny, it is only by living through the undecidability of the unworldly and unhistorical drive to planetary technicity that we will discover the as-yet concealed essence of (digital) being. Undecidability is both a danger and a possibility, just as much as *poeting* technology begins in the language of the unworldly and unhistorical. Who would, in the end, be strong enough to challenge the virtual will by its own gamble of undecidability? to make of the unworldly and unhistorical a new moment of creative intensity?

A Culture of 'Profound Boredom'

> Man himself has become more enigmatic for us. We ask anew: What is man? A transition, a direction, a storm sweeping over our planet, a recurrence or a vexation for the gods? We do not know. Yet we have seen that in the essence of this mysterious being, philosophy happens.[45]

Beyond his understanding of technology in terms of the language of the will to will, Heidegger was, above all, a cultural theorist. Long before its time and cloaking his strictly cultural analysis in a book of

1929/30 lectures titled *The Fundamental Concepts of Metaphysics*, Heidegger was a true visionary of life in the new century – a cultural theorist who laid bare in his writings a culture of 'profound boredom' as the inevitable consequence of the full realization of planetary technicity. Picking up precisely where Nietzsche had stopped before his mind retreated into the silence of madness, Heidegger posed the fateful question concerning what happens 'when man becomes bored with himself.' What happens, that is, when 'temporality temporizes' with such intensity that time becomes a 'drag' on life, when the meaning of dasein (being) suddenly chasms open as an 'emptiness'? Going beyond secondary forms of boredom such as boredom by (situations) and boredom with (persons), Heidegger saw immediately that the abyss awaiting technical consciousness was a sense of boredom so profound, an abandonment of being so generalized, a culture of distraction so pervasive that 'what oppresses us ... is the very absence of any oppressiveness in our Dasein as a whole.'[46]

The absence of an essential oppressiveness in Dasein is the *emptiness as a whole*, so that no one stands with anyone else and no community stands with any other in the rooted unity of essential action. Each and every one of us is a servant of slogans, an adherent to a program, but none is the custodian of the inner greatness of dasein and its necessities. This *being left empty* ultimately resonates in our dasein, its emptiness is the absence of any essential oppressiveness. The *mystery* is lacking in our dasein, and thereby the inner terror that every mystery carries with it and that gives dasein its greatness remains absent. The absence of oppressiveness is what fundamentally oppresses and leaves us most empty, i.e., the *fundamental emptiness that bores us*. The absence of oppressiveness is only apparently hidden; it is rather attested by the very activities with which we busy ourselves in our contemporary restlessness. For all the organizing and program-making and trial and error there is ultimately a smug contentment in not being endangered.[47]

For Heidegger, *we are weak in the ground of our essence*. This is the inevitable consequence of the culture of technicity which, in its deepest functioning, works to make us forgetful of our own being. In the midst of an advertising culture that promotes speeding up the human body to its point of disappearance in the global circuit of consumption and that accelerates the demands of communication until connectivity 'moves faster than thought,' all questions concerning the 'ground of our essence' can be jettisoned as digital detritus. When (electronic) space moves faster than (bodily) time, it is not simply that 'temporality temporizes' but a matter of the disappearance of (embodied) time and (grounded) space into a (digital) 'emptiness that bores' as the very essence of its seduction. Cynical emptiness, then, as the real seduction of digital culture. We live in the 'stretched time' of the *standing now*.

But for all the dynamic momentum of a culture of speed, an economy of networked connectivity, gridded bodies, and algorithmic eyes, there is still the faint whisper of Heidegger, speaking from the memories of the ruins of fascism yet to come, that there is always a heavy price to be paid for weakening of the ground of our essence. Indeed, if we can speak of the present age in the language of virtual fascism, in terms, that is, of virtual technology as a projective form of fascism precisely because it is intent on the radical exteriorization of the human sensorium, then it might also be said that out of this 'profound boredom' with oneself will also emerge new forms of what Nietzsche described as 'monstrous consciousness.' Consequently, the curiosity: to understand the future of the twenty-first century, we turn repeatedly to prescriptions supposedly about the past: Marx's commentary on the commodity-fetishism of the nineteenth century; Nietzsche's prophecies, written at the dawn of the twentieth century, although purposively directed at citizens of the 'next two centuries'; and Heidegger's metaphysics of profound boredom drafted at the beginning of the 1930s.

But then, remember Heidegger: the future is in the past, and that which is close at hand can sometimes only be approached by travelling a great distance. In each case, the gratification for exploring the future of the past is the same: discovering a fundamental *'attunement'* by which to think the present circumstance. While Marx thought capitalism via the attunement of 'alienation,' and Nietzsche *poeted* modernity through 'nihilism,' it was Heidegger's special insight to make of the experience of 'profound boredom' a way of clarifying the full dimensions of the post-human. In its essence profound boredom anticipates the virtualization of human will with such intensity that 'being left empty' is the fundamental *condition of possibility* of digital culture. That the emptying out of being is accompanied by a culture of 'smug contentment' is, in the contemporary situation, undeniable. However, general relief at being liberated, if only by forgetfulness, from the oppressiveness of dasein accompanied by exhilaration at being swept away into the networks of digital connectivity does not mitigate the fact that in this 'emptiness that bores' the worm of being still turns. Ironically, the more distracted the consciousness, the more accelerated the flesh, the more artificial the intelligence, the more centrally pulsed the digital nerve, the more 'standing still' the time, the more proof positive that even the denial of the 'emptiness that bores' confirms the fundamental reality of the *metaphysical* crisis of technological society. That in the hygienic world-picture of digital culture, we never cross the 'danger-zone' of dasein, never 'overreach' the boundaries of being so as to reveal its terror and expansiveness, may account for the very much repressed, yet psychologically active, sense of rootlessness as the essence of digital being. Thus, once again: What is 'being digital'? 'A transition, a direction, a storm sweeping over our planet, a recurrence or a vexation for the gods'?[48]

New Media Art

The paradox of the contemporary situation is that it is only by travelling deeply within the language of political economy of virtual capitalism that we can hope to discover the 'saving power' by which to bring metaphysics, and ourselves, through this period of transition to its next possibility. While life within the accelerated culture of the digital nervous system has the very real effect of situating us within the time of the 'standing now,' of transforming us into manufactured 'raw resources' necessary for the reproduction of the will to will, of sequencing our consciousness with the imperatives of the (technological) object itself, it is only by '*dwelling*' attentively within the digital nervous system that *being* can be forced out of concealment, revealing its possibilities for overcoming the present 'abandonment' of being as the digital future transits between a hyper-nostalgic 'going away' and a virtual future. Networked capitalism, then, as both the dominant business strategy in the era of augmented reality and knowledge-based economies, and the transition to another artistic possibility. Networked capitalism is the most advanced expression possible of the doubled languages of appropriation and revelation.

Understood as a historian of technology rather than as a futurist, that which Heidegger has to tell us about the possible consequences of the fully accomplished society of technicity are already hard-wired reality. Human subjects as the 'standing-reserve'; technological time as the 'standing now'; calculative thought as a 'flight from thinking'; virtual culture as 'injurious neglect of the thing': what Heidegger could only predict has now reconfigured the human sensorium as its digital nerve. While Heidegger could only theorize from afar the image of the 'world picture' of the technological dynamo, today *we* are the world picture. Massaged by the image-machine, motivated by the will to technical consciousness, made restless by the fetishism of money, (digital) labour as fungible and exchangeable nodes with the hyper-realized market-place: what is most striking today is that we no longer live, as did Heidegger, at the advent of technological destining but at a critical moment in its transition to full economic maturity. For this reason, it is our specific destiny to be the first human beings to experience technological consciousness from the inside, to understand the ocular consciousness of electronic culture, as our autochthonous. If we are rootless, then the deserted interior landscape that we populate is the hyper-realism of electronic culture. Technology as destiny is what is closest to us, and, probably for this reason, what is most distant in our comprehension. To be caught up in the destining of technology is a deeply charismatic experience, remarkable for its creativity and inventiveness as much as for its self-confident aggressiveness when (digital) technology and a (new) business class fuse as the basis of a globalized twenty-first century political economy. Indeed, if the essence of technology is coeval with human identity, then it would be churlish to turn our backs on that which is most essential to the question of being. And yet, if we cannot see beyond the glare of digital charisma, if we cannot presence that which is closest to daily life, if we cannot speak of technology in a language of revelation as opposed to appropriation, then our sure and certain fate will be to be reduced to the silence of a manufactured object of technology.

The time is short. The danger is near. The extinction of human beings as a species type is a predictable outcome of technological destining. The will to will brooks no opposition, and it will certainly take its chances with the algorithmic operations of robotic culture rather than with the spectrum of human stubbornness and creative intensity essential to the 'mystery' of being human. In a cruel gesture of the gods, what has been concealed from us in the destining of (technical) being may be the fact that the human project has always been to be a passive vehicle carrying the will to will through its transition to virtual embodiment in a fully autonomous machine culture. A future replicant of chip flesh, algorithmic eyes, and hyper-communications, therefore, as the beginning of future metaphysics, that (virtual) point where the will to will sheds its human skin for a transcendental body of machined genetics and nerve-net flesh.

But, of course, this story, the story of future philosophy, would confuse metaphysics and science fiction, and we know that the canons of technological orthodoxy do not permit such fusion of strange polarities. And yet, if we follow Heidegger's 'pathways,' we cannot be content with simply writing survivors' reports about our enthralling, and no less distressing, fate as citizens of a culture of fully accomplished metaphysics – nodes, that is, within the networks of planetary technicity. Heidegger always thought on a cosmic scale. Refusing the limits of periodicity and overcoming the oblivion of disciplinary boundaries, Heidegger 'presenced' the present by bringing together past and future. He not only thought technology to its origins in the history of metaphysics, but he also wrote out a prolegomenon to a future metaphysics in his deconstructions of the sciences of biology, zoology, and mathematics. Repulsed by 'flight from thinking,' Heidegger brought the aesthetics of incommensurability to bear on the fatal object of its fascination and possible disappearance – the essence of technology. In doing so, all his thought was metaphysics as science fiction, all a story about the constructed reality that is human destiny, from the banishment of the pre-Socratics to the edges of human consciousness to the recombinant logic of quantum physics. When Heidegger proffered that 'being is held out into the nothing,' he did not shrink from the necessary conclusion that radical deconstruction, too, is an element of the human spirit and, consequently, that only a style of being, a 'comportment towards technology,' that could embrace the language of deconstruction while cleaving to the poetics of revelation could prepare us to 'challenge-forth' technology as destiny to its limits of reversibility. Confronted by the double 'challenging-forth' of technological and marketplace determinisms in the era of virtual capital, we desperately need another way of unconcealing the essence of technology, a way of 'thinking' technology that, while versed in the strategies of appropriation, is simultaneously a mode of revelation. The twenty-first century awaits its moment of a Heideggerian 'turning.' It awaits new media art as one way of folding back the language of appropriation into technological revelation.

New media art? That would be Heidegger's concept of the 'turning' as a way of opening up being to the incommensurability of the digital nerve, learning, that is, a 'comportment towards technology' that begins with the premise that the saving-power is also harboured within the danger. While Heidegger articulated the concept of 'turning' within the horizon of mechanical technologies, specifically as a way to recapture the forgotten 'fourfold' of earth and sky, divinities and mortals, the aesthetic possibilities of the 'turning' apply even more forcefully in the context of electronic, digital, and bio-genetic technologies.

For Heidegger, the special purchase of art, understood as a poetics of listening to that 'which withdraws,' is that the artistic imagination recovers thinking in a time of thoughtlessness.

Once we are so related and drawn to what withdraws, we are drawing into what withdraws, into the enigmatic and therefore mutable nearness of its appeal. Whenever man is properly drawing that way, he is thinking – even though he may still be far away from what withdraws, even though the withdrawal may remain as veiled as ever. All through his life and right into his death, Socrates did nothing else than place himself into this draft, this current, and maintain himself in it. This is why he is the purest thinker of the west. This is why he wrote nothing. For anyone who begins to write out of thoughtfulness must inevitably be like those people who run to find refuge from a draft too strong for them. An as yet hidden history still keeps secret why all great western thinkers after Socrates, with their greatness, had to be such fugitives.[49]

Art as fugitive thought, withdrawing into what withdraws, only vital to the extent that it stays within this draft, this current. More than a probe, art is an 'uninterpreted sign,' a pointer, a direction, that works to unconceal that which lies hidden in the essence of technology. While Heidegger reflects on Hölderlin's hymn 'Mnemosyne' ('We are a sign that is not read / We feel no pain, we almost have / Lost our tongue in foreign lands')[50] as a reflection of the enigma of thinking, he might well have been speaking of art, too, as a fugitive 'sign that is not read' maintaining itself in the drift of that which withdraws. Against the current of speed culture, the essence of new media art lies in *reversing* the technological field: an art of (electronic) slowness, an art of digital dirt, an art of boredom. A creative intensity, fugitive art would 'lose its tongue in foreign lands,' becoming the essence of technology with such enigmatic force that the doubled languages of calculation and meditation, appropriation and revelation brush against one another in their passing orbits, creating a third eye of technology: an eye of art that driftworks with the current as it moves forward into the past of that which lies concealed in the language of the technological dynamo.

Is art, then, the pathway opened up, simultaneously a method and an end, by our present transition to the completed stage of metaphysics? And, if this is so, then is the language of art itself the hidden essence of technology, the lost poeisis that since the time of the tragic revolt of Greek rationalism against the original fluxus poets that were the pre-Socratics – Empodocles, Heraclitus, and Aristophanes – has been banished from the ruling vocabulary of 'enframing'? In thinking technology through the medium of art, in meditating upon the new economy through the optics of new media art, do we, however inadvertently, stumble upon a forgotten fissure in the origins of the work of technology? A radical split between techne and art, between 'enframing' and a drift aesthetics that is, at once, the source of the present crisis of technological culture and its possible moment of redress. To reflect on the issue of technology and art is to be drawn into the essence of the 'danger' and the saving-power.' A 'rift' that brings together opposing tendencies into its common outline, Heidegger's perspective was that art is a 'primal strife' that, in its most intense aesthetic moments, is enigmatic because it conceals and unconceals, is sometimes true because it is always untruth. Certainly art is poetry in its essence, but a poetry that dwells in many regions where fugitive thought takes shelter from the storm: in writing, architecture, music, video, computer graphics, multi-media art, performance art, dance, and sometimes even as an art of networked life.

> It is due to art's poetic essence that, in the midst of beings, art breaks open an open place, in whose openness everything is other than usual.[51]

The Future Was Yesterday

Now, thinking thoughts fugitive about Heidegger's vision of art, and, to be honest, not yet convinced that the politics of perfect control that is the digital nervous system can be counteracted by art, particularly when new media art itself is under such pressure to reduce itself to an object of technical consciousness, I took my thinking for a stroll along Pier 39 on the SF waterfront. I knew this was very uncool: a hyper-tourist zone in a city that prides itself on pretending that tourists are only a phantom presence. A city that Kathy Acker once described as a place of 'pooled energy.' But there's something about Pier 39 that is quintessentially end of the continent. Maybe it's the simple fact that the railway to the Pacific literally ends here, or that it's an improvisational free space in which there congregate, like a daily holy mass, all the wild spirits of the street people and the carnival-style hucksters and the disposable-camera tourist crowd and sidewalk restaurants with huge open pots of steaming crab, jammed-together tables, and cell phones and Palms and digital cameras with their human conveyors.

It was one of those radiant March days. The morning fog had lifted, ferries and tall ships and cargo containers destined for eastern lands were working the waters of the Bay, the sun was breaking bright over the Golden Gate, and everywhere I looked there was fugitive art, maybe not swell style like Holderlin's, but art 'break[ing] open an open place,' a fugitive art that wouldn't even dare call itself art. Like the prophet of the streets who is walking backwards down the Embarcadero, worn-out shorts, open shirt, hair balding away the original 1960s style, walking fast and walking slow, all the while jamming out a rhythm song about 'why can't we all learn to live together in the almighty words of Rodney King, beat up by the LA police, but still speaking for the people, by the people, and of the people.' A homeless guy looks his way, and just shrugs his shoulders, disowning all the noise. Two performance artists, silver-sprayed skin, exhausted, draped over their bodies waiting for real time with no place to hide. It's strange to see them this way. Minutes ago, they were the centre of a sprawling crowd: faux robots permanently frozen in position, moving just a slight mechanical levered-motion to acknowledge donations large and small. When they come off their pedestal, their silver robo-costumes which made them look so smooth in the sun suddenly crumple into sad-sack folds, accentuating what most people go through in daily life. To the side, a Latino dancer has set up shop with an unpainted piece of plywood, tap dancing his way to sidewalk cool. But best of all was an African American musician, Onyx Ashanti, or so said the sign by his CD's for sale titled *The Future Was Yesterday*, playing sounds crystal beautiful and deep-blue true on his digital wind sax. Real new media art in the streets of the new century: a laptop for programming, a drum machine for percussion, high-quality microphones wired to a portable, silent generator. *The Future Was Yesterday* could have been playing anywhere, from the streets of Oakland to the finest clubs of American jazz. There was something so perfect in that scene: a fusion of space and sound so evocative that when you turned your eyes from the performance, looking out over the waters of the Bay, past Alcatraz and Sausalito to *The Future Was Yesterday* of all the San Franciscos of the world, well, in that moment, you suddenly found yourself no longer part of the crowd at Pier 39 down on the Embarcadero, but swept up into an open place of art 'in whose openness everything is other than usual.' Just for that speck of passing time, just for that speck of passing art, I suddenly knew what Heidegger meant by the other side of the 'standing now': time folding back on itself, time set aside, and future time – an art of sound and gestures and fatigued performance bodies and springtime sunshine and drifting wind and the end of the continent but the beginning of the Pacific that opens myself up to myself.

HEIDEGGERIAN METAPHYSICS
IN THE STREETS OF CYBERCULTURE

'Injurious Neglect' in Palo Alto

'Lying on the pavement in a public plaza under the overhang of Burger King restaurant in Palo Alto, David Stansfield looked like he was taking a nap this morning.

That was common behavior for the homeless man, known for his nickname, "Doc," his spindly legs, a white beard and the "Gilligan's Island"-style hat that always covered his head.

As shoppers hurried by and diners munched on their burgers and fries, looking through the plate-glass windows onto the plaza, no one noticed that anything was wrong.

No one but another homeless person, who had left Stansfield sitting there an hour earlier and came back at 11:15 a.m., only to find him dead, his face blue, his headphones still covering his ears. He was 54.'[52]

The story of 'Doc' goes on to explain that two nights earlier, he and two other homeless men had been evicted by the local police from the parking garage where they usually slept. Like depression-era railway police but this time at the tech epicentre of the new economy, the Palo Alto cops were as direct as they were brutal. They didn't just roust the homeless men, but took away their sleeping bags and all their possessions. *Disappearing the homeless by abuse value*: precisely what Heidegger meant by 'injurious neglect of the thing' in the mode of ordering involved in willing and doing.

On the same day, another headline reports from Palo Alto on the peculiar problems confronting the 'children of the affluent.' As one seventeen-year old computer entrepreneur who had just earned $75,000 in stock options in a summer job said: 'Most kids here are interested in the market. Because a lot of their dads have become pretty rich and that's why they get to live in a gigantic house.' *Metaphysics on the cyber-boulevard*: the fetishism of money in a hyper-rationalized stock market as the psychological condition of possibility necessary for technological 'destining.' In California these days, there is a new therapeutic crisis involving the children of the leaders of the virtual class: 'sudden wealth syndrome.'

Certainly, these are the stories of two classes in virtual capitalism – one virtual, the other surplus; one ascendant, the other disappeared – but the story of these two classes, or at least of 'Doc' and the seventeen-year old computer wizard, cannot be explained simply economically. On the sunshine coast of California there's a new twist stirring in the pop psychology of the technodrome. Not just physical coercion of the poor by security forces acting at the behest of the affluent, but, in the high-tech communities, a growing hatred of the homeless precisely because they are objects of 'injurious neglect.' The poor are to be punished for being poor, just as much as the homeless are to be vanished for being homeless. For example, in San Francisco, opposition to new homeless shelters in the Mission district comes from a coalition of real estate developers intent on transforming this traditional working-class neighbourhood into a tech business centre and new-economy loft owners who see a future in gentrification. In the housing algorithms of the new economy, the homeless simply don't have a place. They are marginal without a physical margin to call their own. Their bodies are taking up (physical) space in a high-value zone that's suddenly amping up to virtuality. And so, the new economy as a killing field. Not just the death of the homeless man in Palo Alto but the fetishism of dead money as the essence of digital being among the children of the affluent. In the California destiny, the homeless have mutated from an inconvenient fact that won't go away to objects of injurious neglect, whose last pleasuring of the community of affluence is to finally clear the ground by dying, headphones still on, shoppers buzzing by on a sunny Burger King day in Palo Alto. Eyes wide open, the California new economy has something very old as its essence: the defeat of an ethics of caring by the triumph of the technology of abuse value.

Rap Metaphysics in Penal California

It's a rainy afternoon on Market Street in downtown San Francisco, and politics has just broken out.

A large crowd of students, high school age with flying banners, rapping bodies, five-earring ears, and knapsacks stuffed with flyers and poems, have come to the centre of business culture to protest the passage, just the night before, of Proposition 21. First proposed by Pete Wilson, ex–Republican governor of California, Proposition 21 is nominally directed against gang warfare, but, in actuality, goes a long way towards the instant criminalization of the young. In California these days, it's not just immigrants from Latin and South America who are subjected to penal incarceration and forcefully disappeared from the welfare system, but now the panoptic eye of the majority has turned its inquisitorial gaze upon young people. This vast extension of the California security state to harvest the young has a blue chip corporate sponsorship: Chevron, Hilton, TransAmerica, Unocal, and Pacific Gas and Electricity.

A demonstrator hands me a flyer that makes the case succinctly:

This law will:

- make it easier to label and incarcerate young people as gang members on very loose criteria.
- charge non-violent acts as felonies (i.e. graffiti, 'association' and 'conspiracy').
- mix youth with adults in prison and send youth as young as 14 to adult court system rather than to juvenile court.
- force billions of tax dollars away from education and towards building new prisons. (California already leads the nation in prison spending, but is a mere 41st in education spending.)
- NOT provide any funding to prevention programs, despite the fact that prevention programs are estimated to be TWICE as effective and significantly cheaper than youth incarceration (The Rand Corporation).

Not just the homeless, it seems that if you are young in California and born on the wrong side of the virtual tracks, you are likely to be disappeared as well. Raising the spectre of youth gangs running riot in the streets, Proposition 21 facilitates the transfer of young people to adult courts on the recommendation of prosecutors without any interference by the rule of law, specifically by lawyers and judges.

It seems that the new economy has got jailhouse written all over it. Penal California.

I'm standing on the corner watching the demonstration which by now has mutated into a street-smart rap festival, with students – black, white, Chicano, young men and women – taking turns at the mic rapping to an anger-beat about being tied down tight in an education system on the way to nowhere. Someone beside me pulls out a beat-up copy of Ginsberg's *Howl*, and when it is her turn she just stands there reading those verses blind-eyed and bitter true about the 'best of my generation' going crazy in the early morning streets. And I'm taking this all in, thinking thoughts mad and sad about what Heidegger had to say about the 'malice of rage.' How, maybe, the folks outside San Francisco in the Bible belt of the Inland Empire and San Fernando Valley, those sun-baked suburbs running down the spine of California until they take root in the LA desolation sprawl, have been infected with a malice of rage so deep to the psychological core that they are just waiting for scapegoats to appear on the political screen: immigrants for sure, but also the homeless, welfare mothers, and now, it seems, young people in the fading California sunshine. Proposition 21 as the political fallout of 'the emptiness that bores.'

Later that night, I flip on the television and get the breaking news showing images of the very same students I saw in the demonstration being arrested by SF police. It seems that the demonstration got restless on the street, and maybe frustrated with the knowledge that street protests are one thing, but 'entrenched power' quite another, the students took their case to the lobby of the Hilton, demanding payback in the form of an equal financial contribution by Hilton to education as the monies they had paid to sponsor Proposition 21.

I think again about another flyer, this time by STORM (Standing Together to Organize a Revolutionary Movement) which began with the following:

The Set Up

Prop 21 is only the latest shove in a long line of ruling class schemes to push our folks over the edge. They boot families off welfare and in the same heartbeat fire everybody and their momma to 'improve' the economy. Meanwhile they tear down the projects, so the only 'public housing' left is under the freeway overpasses, park benches and door-ways. And to make sure the block stays hot, they flood our communities with a gang of racist robo-cops to terrorize us in the name of 'law and order.' They use these attacks as cattle prods to herd more and more of us into the pen. This pen – the criminal justice system and the prison industrial system – is Massa's new plantation, where inmates labor for slave wages and justice is never served.

Peep game: this system has put the smack down on people of color, women, immigrants, lesbian, gay, bi-sexual and transgender people – politically, economically and physically.

Rap metaphysics: finding the beat of despair and the words of anger, not in what's far away, but what's very close at hand: the prison, the projects, the peep game.

5 In a Future That Is Nietzsche

Heidegger's Nietzsche

Perhaps as a psychological counter-move to his fatal attraction to the stronger will of Nietzsche, Heidegger always insisted that while Nietzsche could brilliantly express the deepest logic of modernist metaphysics, he never succeeded in escaping the axiomatic of his time because of his abiding commitment to the language of value. For Heidegger, Nietzsche's primal concepts – the will to power and eternal recurrence – were themselves simultaneously uncoverings of the modernist episteme and its deepest continuation. Consequently for Heidegger, Nietzsche's story had about it the tragic sense of futility: a thinker who prematurely fell into silence because he had been overcome by the language of completed nihilism – a writer whose thought penetrated the inner soul-ice chamber of nihilism because it had managed to escape the rational prison-house of thought. Beyond writing, beyond thinking, Heidegger's Nietzsche was simultaneously the Antichrist of nihilism and the value-priest that seduces into a virulent life-form the modernist nomenclature.

But then, Heidegger was always in Nietzsche's past. If Nietzsche could begin the preface of *The Will to Power* with the remark that he was writing prophecy for the next two centuries, that would mean that the appearance of Heidegger was already forecast by Nietzsche, that Heidegger's famous deconstruction of being, his unfolding of the will to will to reveal both its 'dangers' and its 'saving-power,' was also part of the Nietzschean matrix. With his failure to address the real meaning, the metaphysical meaning, of Nietzsche's language of value, Heidegger may have thought that he had finally succeeded in leaving Nietzsche behind. But in the fatal gesture of doubled stars, Nietzsche had not yet given his assent to a final break with the Heideggerian rebellion. Could it be that Heidegger is Nietzsche recombinant? A thinker, that is, in the stream of prophetic gesture whose unwitting philosophical role is to complete the cycle of eternal recurrence, to transform the critical modernist form of the will to power into the essential hypermodern expression of the 'will to will'?

And if Heidegger were to reply that unlike Nietzsche's his thought finally dispensed with the value-form of being in favour of the pure nothingness of nihil, that he embraced nothingness as the shadowy spectre that haunts the restless pulsations of the will to power – well, if Heidegger were to make this claim then Nietzsche would perhaps only respond as he once did in his private correspondence that 'I was born posthumously,' that with Heidegger the Nietzschean prophecy of the 'wiping clean of the horizon and the drinking up of the blood of the gods' had its most powerful exponent, that Nietzsche for the twenty-first century was also born posthumously with Heidegger.

With, however, one important exception. At the end of his life, Heidegger might have balanced his desolate vision of technicity as the purest, exterminatory expression of the will to will with a saving-appeal to the incommensurability of poetic utterance. Again ironically, just when Heidegger thought he had escaped the binary language of value, just when he was most confident of transgressing the philological tombstone of the eternal recurrence of the same, he unwittingly worked to reinstall the purest expression of value through the instrumentality of interpretation. After all, what is the dialectic of techne and art, this strange dialectic of 'danger' and 'saving-power,' but the reinstallation of the binary code at a higher level of abstraction and generality, and, may we say it, incommensurability of identity and difference? Yet, it is also something more. In the curious way of the life of thought, where thinking streams from thinker to thinker, from generation to generation, Heidegger's plea for the 'saving-power' of art as a way through and beyond the drive to technicity is also a return to the final moments of Nietzsche's life. It is a return to that moment when, in the late 1890s, the hammer-blow of the pen dropped from Nietzsche's hand, he stopped writing, and in his journey to death first at the insane asylum, then later with his mother and sister, he never spoke again, but would only sit at a piano in an empty room, playing the same notes over and over again. Nietzsche's final silence is the nothingness around which Heidegger's thought hovers, and on behalf of which Heidegger can only write again and again the fatal, irreconcilable signs of identity and difference, being and nothingness, danger and saving-power, techne and art. And who is to say unequivocally that Heidegger's turn to art, particularly to the poetics of Hölderlin, wasn't also in the way of a dropping of the hammer-blow of philosophy from his hand, that Heidegger's increasingly despairing pleas for a politics of incommensurability and a life of undecidability, for the redemptive language of art, wasn't also another desperate gesture by a broken philosopher sitting in an empty room with a piano keyboard for a companion to the grave?

Heidegger repeats Nietzsche, even to the point that he thought that nothingness was somehow different from eternal recurrence and thus an animating source of wonderment and peril. Heidegger thought that he had finally succeeded in substituting paradox for Nietzschean irony. But if Heidegger repeats Nietzsche and Nietzsche was finally broken in spirit by the impossibility of living out in his autobiography a body pulled in the direction of two opposing tension – Apollo and Dionysus – then what are we to do today who live out the final hundred years of Nietzsche's prophecy? Nietzsche's silence is the post-millennial mood. Life as a whirlwind vortex of change, with great historical forces at play, technicity in control, but still the spirit of the body is pulled between two irreconcilable forces, between the virtuality of a space that is publicly crushed by the will to (cynical) power and the materiality of a time that refuses to go backward. What are we to do in a century of Nietzsche's silence, when the sounds of that discordant piano note in an empty room leave their German grave and become the sound-machine of the present century?

The rebellion we face is the eruption again and again of Nietzsche's irreconcilability.

Nietzsche's Heidegger

Nietzsche is the fatal object of attraction and repulsion around which Heidegger's thought hovers like a captive moon. An object of attraction because Nietzsche's reflections on the 'unfolding catastrophe' of human subjectivity following the death of God parallel Heidegger's meditation on the desolation of the human subject, once transformed into a raw subject of technical experimentation. And an object of repulsion because Nietzsche's reduction of the real to a 'perspectival simulacrum' and, with it, the disappearance of the question of being into a theatre of hyper-reality, can never be fully embraced by Heidegger.

While Heidegger shares with Nietzsche a common vision predicated on cultural pessimism, he breaks with him on two decisive points. While Nietzsche disappeared himself in silence rather than consent to pragmatics of the will to truth, Heidegger remained loyal to the possibility of a new regime of anti-truth, a practice of poetics that would represent the first stage in a fundamental 'turning' of the age of fully accomplished metaphysics. While Nietzsche voiced only contempt for those who would philosophize without breaking the table of values, Heidegger sought to save the appearance of metaphysics. Consequently, reading Heidegger and Nietzsche with and against each other is also an exercise in poetry: a counter-poetics of (Nietzschean) madness and (Heideggerian) bitterness. This is a challenge of poetry that Heidegger does not survive. In writing, he might allude to the restorative quality of the poetics of remembrance as an intimation of the withdrawal of the gods from the ruins of the earth, but in life Heidegger is always undone by Nietzsche's counter-challenge of a poetry of the blood. Thus, while Heidegger claims that the fundamental division between the two warring perspectives is that between 'being' and 'becoming,' that Nietzsche's affiliation with the philosophy of becoming sets him in radical opposition to Heidegger's preference for the 'truth of being as being,' this preemptive division between a philosophy of becoming and a metaphysics of being, between Hegel and the pre-Socratics, is itself not a terminal division but a perspectival simulacrum. It is a mirror of philosophy – the necessary 'apparent' opposition installed in the axiomatic of the will to truth, both to make palatable the cycle of eternal recurrence, and to provide the sustaining illusion of transgression as something other than the assimilation of the drive to freedom by the will to truth. Nietzsche recognized in advance the descent of the will to truth into crystal shards of virtuality, but not Heidegger, who sought to preserve the final temptation of the politics of incommensurability. For Nietzsche, the 'apparent' division of being and becoming is the essence of the will to power as a 'perspectival simulacrum.' Always an 'interpretation' – a defining instance of the will to power – the metaphysics of being is the value-directed language necessary to the fully realized metaphysics of modernity. Equally, always an illusion of free subjectivity, the philosophy of becoming is that trace of the wanderer and his shadow that simultaneously finds its preconditions of existence in the orbital mechanics of the language of being, and defines its existence *en soi* in its refusal of the comforting withdrawal of the gods. Thus, Nietzsche's Heidegger is hyper-real: the last affirmation of the truth of metaphysics as the necessary charge to maintain the spin of the atom of becoming. With Heidegger, metaphysics passes into its final, completed stage of being as a mirror of its own disappearance.

If Heidegger is finally trapped in Nietzsche's 'spider web' as its doubled moment of transgression and preservation, then it must also be said that Nietzsche takes refuge in madness against the impossibility of the will to power. Like Bataille after him, Nietzsche might have thought that his was a great rebellion of impossibility against the despotism of the real, that in his thought the 'real' finally vanishes into the aesthetics of simulation, but he was always trapped in advance by Heidegger's warnings about the price exacted for participating in the festival of value. Ironically, in the great sacrificial act of breaking the table of values, in a poetics that was intended to trace out the interior psychology of nihilism, Nietzsche also committed himself to the logic of modernity. Not so much the self-proclaimed Antichrist as the last modernist, Nietzsche was the 'impossibility' that energized the game of utility and confirmed the reality of the hyper-real. It was also Nietzsche's role in the cunning of reason to announce that impossibility most of all is the ruling language of the real.

This is why Nietzsche's writings today can be so influential. Exactly as Heidegger first announced, Nietzsche is the 'fully realized metaphysics of modernity.' This soothsayer of the death of god, this violator of the table of values, this pilgrim on Zarathustra's way, returns as the truth-sayer of the double game of the sign of the real in its final stage of accomplished metaphysics. Read Nietzsche, then, not as a critic from the outside, but as the interior voice of the times. Like Genet's outlaw writing or Blanchot's 'thought from outside' or Bataille's (apparent) rebellion against the real on behalf of a language of incommensurability and paradox, Nietzsche is the enigmatic horizon of the contemporary century. His is the 'accomplished metaphysics of modernity' because the will to power, shedding its rhetorical origins as criticism, has now been installed as the ruling axiomatic of hyper-modern culture. In a strange precession of book culture, each of Nietzsche's texts takes its place today as an essential aspect of the will to power. Consequently, reading Nietzsche always contains a strange mixture of danger, and temptation: an invitation to dissent but also a solicitation to affirm. A crossing-over, a gamble, a precipice, a going across, a danger, and a beginning, Nietzsche is the writer of the doubled sign of meaning.

Not Nietzsche as critic, but something much more dangerous. Nietzsche as 'completed metaphysics,' each of whose books is an assertion of the will to power. That is why Nietzsche is dangerous. He also territorializes the real in the game of incommensurability. Why not think of Nietzsche as also a doubled thinker? A wandering poet whose words flow from him like a prophetic medium. But also Nietzsche as an accomplished metaphysician whose philosophy works to install the language of calculation. Heidegger may not have been able to fully embrace Nietzsche because of his dissent from the language of becoming, but Heidegger was prophetic in recognizing Nietzsche as the realized metaphysics of the times. Consequently, this doubled meaning of Nietzsche, this crossing-over between impossibility and utility, between emptiness and transcendence, is why today Nietzsche is the name we can give to the culture of virtual capital. A culture of virtual capital that simultaneously commands consent as a will to power, but just in that moment of affirmation also slips away into emptiness. Nietzsche, then, as the language of the doubled sign. And this doubled sign is exactly the language of power.

Today, we are all Nietzscheans, trapped in a game set by Nietzsche, a game of power as a flash succession of perspectival simulacra, an aesthetics of political illusions, a will to power revealing itself only by indirection, a power that can speak so affirmatively of the disappearance of power because it is itself a perspectival simulacrum. Maybe his decade of madness at the very beginning of the twentieth century was something other than silence. Perhaps it was a necessary sacrifice, the silencing of Nietzsche as writer precisely because Nietzsche's mind was already authoring a larger, final text – the future of late modernity.

Perhaps this is why reading Nietzsche can solicit such 'profound boredom' in Heidegger's sense. Perhaps reading Nietzsche is also a way of coming to terms with what Heidegger described as 'the ultimate distress as the lack of distress,' that Nietzsche can be so nostalgic for the return of the gods because he already sensed the coming of a culture where everything recycles in recombinant form: cloner bodies, sequenced brands, transgenic fashions, data flesh. When Nietzsche becomes the body of hyper-modernity, when Nietzsche becomes a 'creative value' overturning the table of (modernist) values, then suddenly 'thought from the outside' reverses itself, becoming the transgression that is to be assimilated as part of the language of the will to power. The fundamental transition between modernity and hyper-modernity, Nietzsche's fate is to be transformed into what he thought he was only describing – a going-over, a gamble, a rupture of such magnitude and originality that the gods stir themselves and take notice.

We, who live in the future forecast by Nietzsche, do not know how long this transition period will last or what the final outcome of the transition will be: the appearance of recombinant life-forms programmed with software intelligence, the slow decline of the human species into prosthetically enabled post-humans, or episodic, violent bouts of fantastic explosions of creative energy and fatal implosions of political and social regression? Living within the embrace of his philosophy, we can only know this for certain. The future of any historical eschatology is always a playing-out of its axiomatic. Consequently, if Nietzsche is the fully realized metaphysics of this time, if is difficult to comprehend Nietzsche as the will to truth of a society that he both prophesied and transgressed, then it can only be by taking Nietzsche by surprise, making of his texts a 'creative value' that works to reverse the direction of his genealogy, that we awaken to the Nietzschean character of the times. He is not in our past, but in our future.

Desert Hermit

Is it possible, just possible, that the triumph of the digital nerve is the predestined outcome of what Nietzsche once described as the 'will to nothingness'? And if this is the case, then what is to be our fate in a universe where the will to will finally flees its abode in the sepulchre of human flesh, travelling at light-speed and light-time across the networks of electro-culture, here nesting in the game-nodes of human imagination, there working to install the digitally networked world body of virtual capital, sometimes parasiting the residual remains of industrial culture, always speaking enthusiastically in the language of an indeterminately expanding future?

What is the fate of Nietzsche today? Is his fate to be the writer of increasingly desperate, increasingly lyrical books abandoned to the drying winds of the digital matrix, meditations on the fate of modern being desiccated by the triumphant logic of nihilism? And is there something in the relentless anxiety of Nietzsche's thought, in the pilgrimage of Zarathustra across the horizon that has been wiped clean by the myth-destroying logic of rationality, that makes of his thought both the final confession of a thinker who refuses to cede to the rising sun of the technological times and the first intimation of a style of thinking that, suspecting the ferocity of the coming storm of nihilism, steels itself to the point of resigned silence and implacable hermeticism as a way of strengthening thought for the approaching digital apocalypse? Why in the face of the 'will to nothingness' did Nietzsche choose solitude? Why when confronted by the psychological after-shocks of the death of god, which is to say the death of the ideal of the transcendent in human affairs, did Nietzsche purposively turn to a very (early) Christian response – the life of a hermit? While once other hermits in Africa and Egypt withdrew to the desert as a way of stripping away the temptations of the flesh and preparing for the coming of the messiah, Nietzsche's withdrawal into the silence of the life of a philosophical hermit was also a way of disciplining the mind as he awaited the exit of the Christian god. In a strange inverse spiral of bodily preparation, Nietzsche prepared himself, and perhaps by example ourselves as well, for a method of thought – intense, lyrical, poetic, uncompromising: mythic philosophy – and a method of living – solitary existence – experienced to its depths without being spiritually crushed by the tempests of the will to nothingness. Nietzsche was the first desert hermit of the digital age.

Digital Ressentiment

This is supposedly the long-awaited 'age of communication' and 'accelerated culture,' an era in which even the consumer appliances of everyday life get into the speech act, beginning to whisper telematic commands to one another, a time of 'smart machines' and speed bodies and wireless minds and 'connected eyeballs' and streamed images. It is a time in which the externalization of consciousness so necessary to Internet consciousness – the will to communicate – has about it such a degree of obsession and dynamic cultural foregrounding that its very obsessiveness, its heightened degree of almost-mandatory compulsiveness and open-ended futurism, opens up the question of the 'why?' The question of 'why digitality?' and, particularly, 'why in this culture?' cannot be answered satisfactorily within the ruling axiomatic. It is in the nature of a governing axiomatic to guard itself, stolidly and well, against the undermining of its logic by providing in advance an array of digitally satisfying responses to the question of the why. A blindingly brilliant digital future opens to the initiated. But for those recalcitrant minds who thought-drift back to the foundational query of 'why digitality?' and 'why now?' it might be to the preparatory thought of a desert hermit that one would turn for an undermining reflection. In *Thus Spake Zarathustra*, Nietzsche said that the ultimate frustration of the dynamic will is that 'it cannot turn time backwards,' that the will to will is ultimately frustrated by the passing of 'time's it was.' As digital reality, this successor to the exiting of the Christian God, projects itself forward in the accelerated language of light-time and light-space, as the will to universal space defeats, indeed humiliates, the reality of particular time, might there not also be heard in the command language of digital futurism a perceptible hint of ressentiment? Could it be that digital futurism for all of its technological wonderment, for all of its bountiful expression of the will to communicate, has about it the familiar scent of revenge-taking? In an ironic turn in the logic of eternal recurrence, isn't the present era somehow similar to other periods of technical enlightenment in which the dull implacability of 'time's it was' is supposedly vanquished by the will to time-binding light-space? A culture of digital ressentiment.

But, of course, in the culture of advanced technicity, at that precise historical juncture wherein software codes combine with genetic engineering and molecular biology to produce giddy visions of bio-chips recombinant, we have supposedly abandoned the sacrificial language of revenge-taking. Ours is a culture ruled by the illusion that we have finally resolved the irreconcilabilities of time and space that drove Nietzsche to the desert spaces of his mind. But, what if we haven't? What if the exuberant rhetoric of digital futurism can't hold? What if the speed culture of streamed visions and flash noise can be so dynamic precisely because its existence is predicated on a repressed amnesia of a more fundamental paradox in the human condition? And what if in a not untypical response to amnesic repression, the culture of digital reality is a global repetition of patterns of individual psychopathology? In this case, the drive to the light-time and light-space of digitality would also be in the way of revenge-taking, a violent experiment conducted on a planetary scale to finally respond to Nietzsche's insight that the last frustration of the will to will is its inability to turn back the stone of 'time's it was.' Isn't this exactly what Heidegger meant by a culture of 'profound boredom' – a split culture where the passing of time is repressed in favour of the will to (spatial) nothingness?

Today, the will to power speaks in the language of the digital nerve. An expression of (techno) life itself, the digital nerve is a violent force-field, pulsating with energy, netting with a utopian will to connectivity, vibrant with auto-emotion, and glowing wireless with streamed data. Disconnected from the ancient gods that inspired the will to (sacrificial) power of antiquity and abandoning open loyalty to Christian theodicy, the digital nerve is that point where the will folds back upon itself, becoming in the form of the will to will its own grounds of justification and ultimate goal. A will-less will, the digital nerve operates in the form of the mirror of virtuality. Here, the fatal tensions of human flesh in the world, the immediate irreconcilabilities of time and space, are transcended in favour of a digital logic that mediates the after-images of light-space and light-time. What the French philosopher Teilhard de Chardin once described in religious terms as the electronic 'noosphere' has now been realized in the relentless pragmatic, soft engineering language of the electric eyeball. It is as if life itself has disappeared into the optical nerve of the digital eye, and a pure circulation of optically refracted digital data has substituted itself for the polar tensions of human flesh on its way to a meeting with 'time's it was' and human space as somehow integral to questions of memory and social vision. When the digital eye is ripped from the dark cavity of the skull, when the optical nerve is externalized in the transcendent form of streamed data flesh, then the only sound to be heard may be the discordant playing of a piano keyboard in an empty room. And why? Because the abiding language of myth will not be denied. The more extreme the externalization of consciousness in the form of digital media, the more violent the 'eyeballing' of society in streamed data culture, the more confident the technical harvesting of the human remainder, the more perceptible the keening sound of something future-lost in a spatially over-determined digital nerve. Just when we thought we had finally overcome Nietzsche, we merge with the digital nerve only to find the retelling of a more ancient account of technological hubris. Nihilism haunts digitality.

Nietzsche as 'Completed Metaphysics'

Nietzsche is a cipher, a philosopher who might have lived in the dying shadows of the nineteenth century but whose nervous system tapped directly into the future vector lines of the will to power. Literally fast-forwarded by his consciousness of the nervous system of power, Nietzsche's body was burned up by thought. That is why he would walk ten hours a day, take opium to settle down his persisting 'inner tensions,' speed-write in a waking delirium of words, poet lyrically and philosophically and madly but then compulsively write the same letter to friend after friend telling the story of life's platitudes: what he ate for dinner, the view from his room, gestures made by waiters, sounds from the street, bills to be paid, publishers' printing errors, citations of the name Nietzsche in Denmark, France, Russia, and maybe even Germany. Always a doubled thinker, Nietzsche could speak beautifully of the ecstasy of the 'dancing star' but complain bitterly of a never forgotten rosary of personal slights. Perhaps more than other writers, Nietzsche was literally written to death, 'born posthumously' because his fate was to be a word machine, a memory machine, a futurist machine of the will to power. He was a perfect oracular candidate because between his personal psychology of bitterness and his prophetic cast of mind, between churlishness and searing aphorisms, Nietzsche played out the doubled game of power in the flesh. Himself an assimilator, an appropriator, an aggressive conqueror with an overwhelming desire to substitute his name for Christ's, a will to power in the (philosophical) flesh, Nietzsche was also, and simultaneously, a lonely hermit, a solitary thinker, dream-writing *Thus Spake Zarathustra* while chewing on old bones of resentment for moral nourishment.

As the writer of incommensurability, Nietzsche can only be read ambivalently. Try to capture one crystal shard of his thought, to finally materialize what he has to say about 'ressentiment' and 'blond beasts' and 'ascetic priests' and the surge economy of nihilism, and there glancing from the refracted mirror suddenly appears the opposite face of Nietzsche – the hyper-modern thinker who deploys a critique of the language of nihilism only to project himself forward as the leading 'ascetic priest' of philosophy. In ancient times when the dead skull of technicity had not yet cleansed the world of the mysteries of mythology, when hybridities of animals and gods and humans walked the earth, Nietzsche would have been recognized for what he always was, and still is: the figure of a minotaur, half-man/half-god; half-flesh/half-machine – a thinker of degree-zero, a site of intense contradictions where all the differences meet, and are all the more energized. Indeed, could it be that with the abandonment of the desolation of the earth by Heidegger's gods, Nietzsche was one of the alien elect, a fast-decaying opening of skin and nerves and words and thoughts and loneliness, in short, an abandoned thinker, chosen to set in play again the favourite game of the gods – the doubled game of power? And if this is so, then could it also be that the name of Nietzsche is also the doubled gesture by which the gods leave the traces of a sustaining doubt that has not been erased, and will not be erased, by the culture of technicity?

As the favourite son of the now-debunked gods of ancient mythology, Nietzsche fully embodied the metaphysics of the will to power. Nietzsche is completed metaphysics. If he could proclaim in *The Will to Power* that 'interpretation is power,' then that was because Nietzsche only theorized that which he had already experienced autobiographically. Himself the *übermensch* of power – an incessant 'interpreter' of morality, power, sexuality, economy, philosophy, and civic morality and religion –

Nietzsche was of the rare order of thinkers energized to such a point of delirious excess that they sought to assimilate their intended objects in a vertigo of language. Nietzsche had two wills, a will to power and a will to dance the stars. In the specific resistances of these opposing wills, in the daily tensions of embodying the irreconcilable tensions of existences past, future, and present, he was the representative figure of the 'next two hundred years' after the close of the nineteenth century. A split thinker in a split culture, Nietzsche is ultimately trustworthy because he was in the most fundamental sense 'human, all-too-(post)human.' Alternatively perceptive and prejudiced, projective and retrospective, critical and dogmatic, free-spirited yet self-hating, Nietzsche is the first sustained appearance in the modern era of human desolation reflecting on itself. Accidented by the will to power, Nietzsche's mind is a grisly report on the age of completed metaphysics that is still in the process of fully disclosing itself. In his own words, Nietzsche is simultaneously the 'overman' and the 'no-man.'

A Double Death

This is why *On the Genealogy of Morals* retains its power as a sustained deconstruction of contemporary society. This is a text that can only be read invisibly, less for what it describes than for an impossibility that hovers at the edge of the text like a recursive horizon – a phantom projection that is all the more insistent because of its unarticulated silences. Ostensibly about the death of god and ressentiment and bad conscience and guilt, *On the Genealogy of Morals* only becomes truly interesting when it is read projectively. Because what is projected here is a double death: first, the fate of modernity living within the shadow of the death of god; and then, the fate of hyper-modernity, the contemporary era, under the sign of the death of man.

Writing at the end of the nineteenth century, Nietzsche claimed that he was a futurist of the next two centuries. About this he is correct. Rather than be read retrospectively as a genealogy of the origins of morality in human anthropology, Nietzsche's thought forces the future to reveal its hidden metaphysics by the method of genealogy. This is a book which travels backward in time in order to fast-forward the future. If, as Baudrillard claims in his reflections on photography, new technologies always begin with the disappearance of the real, then it might also be said that the contemporary convergence of digital technology and bio-genetic engineering also begins with a fundamental disappearance – the death of man as a precondition for the triumph of the culture of pure technicity with its futurist scenarios of nano-technology, recombinant genetics, wireless brains, stem cell research, cloning, speed bodies, and wireless communication. If today we should inquire into the broader cultural conditions underlying the seduction of wireless communication and the quick-time transformation of the human species itself into an object of recombinant experimentation, then perhaps it is to Nietzsche's genealogical method that we should look for a compelling response, a response that finds in the technological drive to fully accomplished metaphysics a substitution-effect for the double death that is the essential precondition of the culture of technicity.

This is why Nietzsche's *Genealogy* should be read mythologically. While the contemporary drive to a technologically overdetermined society imposes silently, but no less pervasively, a linear view of history – the story of the will to power as an always relentless forward motion on the edge of novelty and experimental newness – Nietzsche's lesson is that history always comes too late for the story. Read the *Genealogy* in reverse, rub the utopian story of digital connectivity against the more ancient tale of the origin of morality in terms of the equivalence of pain and punishment, and what emerges from the mists of the past is a strikingly contemporary account of technological values. What Nietzsche described as the 'pleasure of cruelty' that attended the great *auto-da-fes* with their ritualistic presentation of the spectacle of public punishment, this descent of justice into a primitivism of blood equivalence – a political materialism of bodily pain – is the present and future of fully realized technological society.

What will succeed Christian nihilism at the end of Nietzsche's *Genealogy* has always been one of the great cultural mysteries. Written today, would the *Genealogy* be compelled to conclude with an essay on artificial flesh and electric eyes and robotic intelligence – a transhuman legacy for all the bundled ressentiment and repressed fury of the 'caged animal' that Nietzsche foresaw as the majority citizen of the new world? Or is the final chapter of the *Genealogy* already written in the words of the opening chapter? Not a dramatic movement towards a cleansed technological future, but something else – a fatal pairing of the instrumentalities of the drive to techncity with moral primitivism? In this case, why should Nietzsche's books be relieved of the possibility of the 'creative value' that he proclaimed in *Thus Spake Zarathustra* – a 'creative value' that energizes the will to power by reversing the direction of the table of social values, overcoming dogma and sedimentary knowledge? Read Nietzsche, then, against Nietzsche. Allow the *Genealogy*, to be its own 'overcoming.' In which case, the future of ascetic ideals prophesied by Nietzsche is suddenly animated by a new moral eugenics of pain and punishment: a return in hyper-technological form of the classical rituals of scapegoating and sacrifice and the joys of cruelty and the cynical pleasures of inflicting pain and pure revenge-taking.

The Order of Values

This is why the *Genealogy* is such a superb guide to understanding virtual capitalism. Marx may have correctly diagnosed the resolution of the flows of capital into the pure speed of virtual exchange. Heidegger may have written eloquently about the planetary death drive of technicity accompanied by the historical amnesia of profound boredom. But Nietzsche is different. Working in the quiet and unnoticed 'grey matter' of genealogical research, he has uncovered a century after his time the hidden moral axiomatic of virtual capitalism. Refusing in advance Foucault's question in *The Archeology of Knowledge* as to what conditions of knowledge both horizoned and made possible the prevailing consciousness of modernity, Nietzsche is more fundamental. In all of his books, but particularly in the *Genealogy*, he problematizes the question of value itself. For Nietzsche, the basic question that both undermines and propels the metaphysics of the will to power is what order of values makes possible life in an increasingly technical age. To live as cybernetic beings, part digital/part flesh, in an 'adventure' that demands enthusiastic assent to the disappearance of the real, what order of values must silently be set in place as its fundamental precondition? About this, Nietzsche is adamant. Not just the content of values, for that shifts radically as history mutates around the sun of Christianity, but the *form of valuing* itself must be made problematic if its reign as the ruling axiomatic is to be challenged.

Scanner Nietzsche

> I walk among men as among fragments of the future: of that future which I scan.[1]

A critique of the form of values? That's the essence of Nietzsche's contribution to an understanding of the genealogy of digital morals. Now that we live a century after the writing of Nietzsche's prophecies, one hundred years into the future of *The Twilight of the Idols* and *Ecce Homo* and *The Will to Power* and *Thus Spake Zarathustra* and *On the Genealogy of Morals*, we can finally see Nietzsche for what he was: a chronicler of the vicissitudes of human experience, a street philosopher of the last man of Christianity, whose importance would not lie in the past but in the future as the preparatory moral condition – the clearing of the ground of human flesh – that was necessary for the triumph of the digital nerve. That is why all of Nietzsche's writings have about them a tremulous sense of expectation, a feeling beyond normal space and linear time that each text is about to enter into a third dimension of nihilism, that everything which has happened up to the beginning of the twentieth century – the creation of the last man of Christianity, will-less and 'weary' of himself, drifting in history as a moral vacancy waiting to be defined, to be given a value-direction – that all of this has no definitive meaning in itself, but only as the moral context out of which will emerge the fully realized society of technicity. Nietzsche is a historian of defeated and humiliated subjectivity, the really existent material moral condition out of which will emerge the virtual will.

Read retrospectively, *On the Genealogy of Morals* can have no other meaning. It is a mythic story not only of the creation of humiliated subjectivity, created first by the birth of God and then abandoned to its own purposes by the death of God, but a theorization of the moral eugenics by which 'a will was burned into man.' Ironically, it may well be that Christianity and digital technology are deeply entwined, that both emerged out of the same dynamic drive to make of man a will and nothing but a will, reducing the material of human flesh to a secondary sign of the success or failure of the moral eugenics necessary for the fulfilment of the triumph of the will. Now that the theocratic cosmology of Christianity has been blown away by the desiccating winds of secular culture, now that the sacred space foretold by the Old Testament of the name of God has itself been undermined by a cynical drive to simulate to distraction the moral right to give names, might not it also be said that perhaps the will to Christianity and digitality were always flip sides of the same historical movement, that Christianity was always a sustained period of moral preparation for the coming to be of the digital nerve? Today, the mask of Christianity is removed, only to reveal the triumph of the digital gods.

For Nietzsche, the lasting cultural significance of Christianity has little to do with its liturgical significance as a particular act of faith, and everything to do with the fact that under the sign of Christianity a particular order of values was installed in human flesh. A carrier of a dominant cultural meme, Christianity was the method of moral eugenics by which a sovereign self was constructed, representing both an embodiment of the ascetic ideal and a certain sign of the defeat of the vicissitudes of human flesh. In Nietzsche's judgment, we are always born at the precipice of defeat. Of course, what Nietzsche could not foresee is that he himself was also the unwitting agent of the continuation of the story of the will, that the name of Nietzsche may have been the 'gamble' by which liturgical flesh of the Christian will could finally be dropped in order to reveal a future of pure will – the drive to planetary technicity. In this case, the importance of Nietzsche may be as a point of fatal turning, that moment when the sacred object of the (Christian) will recedes from view, only to reveal what has always been the abiding truth of the will to will, that the humiliated subjects – the 'sovereign individual' – would prefer to will nothingness rather than not will at all. In this sense, the triumph of the digital nerve may be, in fact, the real beginning of the age of Christianity, the return of the sacred object signified by the name of God in the form of a will to nothingness that is inspiring because of its depthless emptiness, its 'aimlessness' as it sky-drifts across the horizon of social events.

Christianity as a preparation for the will to digitality? Or the digital nerve as the precondition for the resurrection-effect of a renewed Christianity? Or both? The cynical will of the digital nerve and the bad conscience of Christian flesh as necessarily twin impulses, the tension of which can be so animating because it is all in the nature of a single spiral of virtuality. Here, the creation of the sovereign individual is the moral axiomatic of the digital future, and the digital nerve is the fully realized space of contemporary religiosity. This is why perhaps the animating mythology of classical Christianity and the ideological hype of digitality share a common, abiding vision based in a hatred of human flesh, and a transcendental urge to escape the mortality of the body in favour of a telic destiny, sometimes under the name of God and, now, under the sign of the electronic noosphere of the wireless future and software intelligence and telepresencing as human destiny. It is also why digitality has about it a panic quality of mythological regression, a turning backward in a speed spiral of descent to the originating moments of moral eugenics. Now more than ever, the Christian past is the digital future. This is the secret which Nietzsche did not take to his grave, but wrote about in blood as the sustaining prophecy for the next two hundred years of 'conscience-vivisectioning' and the hubris of 'experimentalism.' For him, we are now, as we have always been, digital 'nutcrackers of the soul.'

THE AXIOMATIC OF MORAL EUGENICS

Streaming Flesh as the Mnemotechnics of the Data Body

> If something is to stay in the memory it must be burned in: only that which never ceases to hurt stays in the memory. That is the oldest psychology on earth. Man could never do without blood, torture, and sacrifices when he felt the need to create a memory for himself – pain is the most powerful aid to mnemonics.[2]

The sustaining myth of technological culture is its belief in the autonomy of the sovereign individual: the individual as the repository of a free will with a right to make promises; the sovereign individual as the possessor of rights and responsibilities; the increasingly neo-liberal self with a right to economic acquisitiveness. The possessive individual who if he or she becomes gradually and imperceptibly 'possessed' by the brand icons of the globalized marketplace still performs the necessary rhetorical function of being the agent of independent social choice in a marketplace that is proclaimed as breeding an increasing cornucopia of economic abundance.

However, Nietzsche has a different story to tell about the origin of the sovereign individual, the political essence of the state, about what is really involved in setting upon the stage of human affairs an individual who is 'regular, calculable, uniform' with the right to make promises.[3] Both classical and contemporary liberalism may provide us with a vision of individualism as the spearhead of the dynamic will, making and remaking the world, an agent of boundless freedom limited only to the constraints of the competitive marketplace, a possessor of a will, a possessor of his measure of value. Not Nietzsche. He argues that what we witness now in the rhetoric of the sovereign individual and the pluralistic state is the product of a huge apparatus of psychic repression: a mode of repression so deeply enmeshed in the political horizon that it has buried itself inside human memory as the will. For Nietzsche, 'breed[ing] an animal with the right to make promises'[4] also implies that 'forgetfulness is the basis of the modern self.'[5] For him, we are twenty-first century products of a long history of moral eugenics, enclosed within the perspectival illusion of the sovereign individual, gnawing at the bars of this rhetorical tomb like chained animals, wasted synapses in a consumer machinery of pseudo-choice, but sometimes also violently 'accidenting' the sovereign individual in acts of objectless rage: road rage, air rage, job rage, sex rage, life rage. For every lone gunman on the city streets initiating another media cycle of vicarious shock and cynical condemnation there is Nietzsche reading out the script in advance:

> Now this animal which needs to be forgetful, in which forgetting represents a force, a form of robust health, has bred in himself an opposing force – memory, with the aid of which forgetfulness is abrogated in some cases – namely those where promises are made.[6]

> Man himself must first of all have become calculable, regular, necessary even in his own image of himself, if he is to be able to stand security for his own future, which is what one who promises does.[7]

> This precisely is the long story of how responsibility originates. The task of breeding an animal with the right to make promises evidently embraces and presupposes as a preparatory task that one first makes men to a certain extent necessary, uniform, like among like, regular and consequently calculable. The tremendous labor of which I have called the 'morality of mores.'[8]

The dominating instinct is 'conscience.' It is burned into the will. It rebels against contraventions of the code, but it is seduced by the language of transgression. It speaks in the language of guilt and bad conscience, and its origins are always tribal, 'soaked in blood.'

> Everything great is soaked in blood for a while. We come out of a history of cruelty.[9]

> To see others suffer does one good, to make others suffer even more: this is a hard saying but an ancient, mighty, human, all-too-human principle. In devising cruelties, they anticipate man and are, as it were, his prelude. Without cruelty, there is no festival: thus the longest and most ancient part

of human history teaches – and in punishment there is so much that is festive.[10]

What is repressed in the language of technotopia is the origins of morality in the history of cruelty. Cruelty to ourselves in burning a dominating will in the form of a conscience into human skin, and cruelty to others in the original materialism of morality where settling a debt relationship involved a pleasure exchange – the right to vent 'his power freely upon one who is powerless, the enjoyment of violation, voluptuous pleasure. The punishment is a warrant for cruelty.'[11]

What we witness today is a liberation of technological beings from this long history of repression. The festival of cruelty, of voluptuous pleasure, returns to public life, sometimes in the form of public policies of the austerity state that aim directly at the powerless, the poor and the homeless, sometimes in the spectacles of the war machine that initiate virtual wars with televised scenes of human suffering as a video exchange of mass pleasure for political debt, sometimes in a moral language of disappearances that literally vanishes the so-called 'losers' of the globalized marketplace from individual conscience. Because what's really at stake is the conscience. Unable to bear the pain of the memory which is burned into it, the conscience, this 'dominating instinct,' turns outward for satisfaction. The 'good conscience' looks for victims, for likely scapegoats, upon which to vent its displeasure. If Nietzsche can murmur from the sidelines 'someone must pay for my feeling ill,' it is because the 'dominating instinct' has become the ruling tendency of contemporary culture. The 'good conscience' finally breaks the silence of religious repression only to go tribal immediately, to participate in great orgies of public cruelty. Providing social scapegoats to appease the violence of the dominating instinct of the bad conscience, that's the rhetoric of public life in the culture of fully realized technicity. Punished if it transgresses the business logic of the digital nerve, always speeded-up, stressed out, over-managed, over-digitalized, over-determined, face busting with anxiety, body treadmilling with instinctive competitiveness, data-mined, digitally netted, flow charted, wireless communicated, the sovereign (digital) individual is all (data) memory burned into the will. It feels digitally ill. It breaks wide open under the pressure. It wants to disappear. It does disappear. But it always finds itself just in (digital) time. And why? Because the digital self is the essence of mnemotechnics: the first axiomatic of the moral eugenics necessary for accomplished technicity. In a culture dominated by information technology, the 'sovereign individual' must either align itself with the main vector lines of the digital nerve, or be trashed as human remainder. Here, the (digital) memory that is burned into will is of a possible future interface of flesh and software, of transcending the body into the circulating tissue of the digital nerve. Streaming flesh as the mnemotechnics of the data body.

'Nutcrackers of the (Electronic) Soul'

> ... our whole attitude towards nature, the way we violate her with machines and the heedless inventiveness of our technicians and engineers, is hubris; our attitude towards God as some alleged spider of purpose and morality ... is hubris; our attitude towards ourselves is hubris, for we experiment with ourselves in a way we would never permit ourselves to experiment with animals and, carried away with curiosity, we cheerfully vivisect our 'souls'; what is the 'salvation' of the soul to us today? Afterward we cure ourselves: sickness is instructive, we have no doubt of that, even more instructive than health – those who make sick even more necessary to us today than any medicine men or 'saviors. We violate ourselves nowadays, no doubt of it, we nutcrackers of the soul, ever questioning and questionable, as if life were nothing but cracking nuts; and thus we are bound to grow day-by-day more questionable, worthier of asking questions, perhaps also worthier – of living?[12]

The essential precondition for the fatal merger of information technology, biogenetics, and the new economy is a relentless spirit of radical experimentalism applied to the soft tissue of data flesh. In the business culture of the new economy, the products of this adventure in radical experimentalism applied to willing human subjects are called 'eyeball culture.' That is, the disappearance of the human subject into the externalized sensorium of the digital nervous system that can be monitored and counted and merged and massaged and manipulated and streamed and resequenced. Indeed, if commercial new media sites today seek so desperately to share bragging rights in how many 'eyeballs' they can deliver to brand advertisers, this is not simply an exercise in the commodification of digital culture, but something much more profound and in that profundity desolate. It is a matter of psycho-anthropology: the harvesting of digital flesh into specialized, distended organs of sense perception. *Alienated vision.* Here, the unitary human subject is broken down into its (electronically) alienated sensoria: some all a digital eye, a cyber-ear, an artificial taste, a telepresenced touch, a simskin smell. The masters of the new economy, then, as so necessary because they perform the instructive role of 'those who make sick,' and the spirit of technological experimentalism as the creative heir of 'two centuries of conscience vivisection[ing] and the self-torture of millennia.'

Digital Bad Conscience

If the essence of the will to technology is the spirit of radical experimentalism, then the psychological motive for 'two centuries of conscience-vivisection' is the installation of the 'bad conscience' as the controlling mechanism of increasingly technical beings. About this, Nietzsche is explicit.

> I regard the bad conscience as the serious illness that man was bound to contract under the stress of the most fundamental change he ever experienced – that change that occurred when he found himself finally enclosed within the walls of society and of peace. The situation that faced sea animals when they were compelled to become land animals or perish was the same as that which faced these semi-animals, well adapted to the wilderness, to war, to prowling, to adventure: suddenly all their instincts were devalued and 'suspended.' From now on they had to walk on their feet and 'bear themselves' – a dreadful heaviness lay upon them. They felt unable to cope with the simplest undertakings; in this new world they no longer possessed their former guides, their regulating, unconscious and infallible drives; they were reduced to thinking, inferring, reckoning, coordinating cause and effect, they were reduced to their 'consciousness' – their weakest organ.[13]

> All instincts that do not discharge themselves outwardly turn inward – that is what I call the internalization of man: thus it was that man first developed what was later called his 'soul.' The entire inner world, originally as thin as if it were stretched between two membranes, expanded and extended itself, acquired depth, breadth and height, in the same measure as outward discharge was prohibited.[14]

So, Nietzsche begins with philosophical anthropology, with a vision of humans voluntarily shutting themselves up in the cage of society; thereby providing themselves with no way of discharging their instinctive drives outward. From now on, the 'instincts of wild, free, prowling man' turned backward against man himself. Hostility, cruelty, joy in persecuting, in attacking, in change, in destruction – all this turned against the possessors of such instincts: that is the origin of bad conscience. For Nietzsche, we are born guilty, guilty of 'being human, all-too-human.'

Consequently, this searing vision of damaged human psychology emerges:

> The man who, from lack of external enemies and resistances and forcibly confined to the oppressive narrowness and punctiliousness of custom, impatiently lacerated, persecuted, gnawed at, assaulted, and maltreated himself; this animal that rubbed itself raw against the bars of its cage as one tried to 'tame' it; this deprived creature, racked with homesickness for the wild, who had to turn himself into an adventure, a torture chamber, an uncertain and dangerous wilderness – this fool, this yearning and desperate prisoner became the inventor of the 'bad conscience.' But thus began the gravest and uncanniest illness, from which humanity has not yet recovered; man's suffering of man, of himself – the result of a forcible sundering from his animal past.[15]

But remember, Nietzsche always speaks in double meanings: This sudden change of state may be less an adaptation than an ineluctable disaster, but still the gods are interested because human beings 'now give rise to an interest, a tension, a hope, almost a certainty, as if with him something were announcing and preparing itself, as if man were not a goal but only a way, an episode, a bridge, a great promise.'[16]

One hundred years after Nietzsche's prophecy, it might be said that the morality which began with the 'instinct for freedom' turning against itself in the form of bad conscience now effects a second 'sundering.' Not the separation of the human species from its animal past, but the separation of the will to technology from the human species. Not the heaviness of gravity as humanity crawls out of the primordial mass of the ocean, but the lightness of digital culture as technical beings – half-flesh/half-node – slip into the skin of electronic culture. Not the 'reduction to consciousness – the weakest organ' – but post-humans driven onwards by the externalization of consciousness in the form of networked intelligence and sim/entertainment and cloner brains. Not the 'internalization of man' as drives are forced to project themselves inward, but the exteriorization of the drive to planetary technicity as we come to live inside the noosphere of the electronic soul – 'the entire external world, originally as [digitally] thin as if it were stretched between two membranes, expanded and extended itself, acquired depth, breadth, height, speed, connectivity, and extensiveness, in the same measure as outward discharge was encouraged.' Not 'homesickness for the wild,' but the abandonment of human flesh in favour of a deep longing for the fulfilment of its telic destiny. Not the interface of the human sensorium with the digital nerve, but the projection outwards of bad conscience itself.

Abandoning its nesting place in human consciousness, the bad conscience goes electronic. Digital flesh – this 'adventure, torture chamber, wilderness, homesickness' – can no longer project its drives inward, can no longer feed on the ressentiment of human flesh, can no longer chew on the chestnuts of human resentment, can no longer parasite itself on the alibi of man as a 'chained animal.' The 'end of man' also means the 'end of ressentiment' as the energizing force of history; the end of the myth of bad conscience indicates the end of the psychology of repressed internalized drives. Man has been overcome; internalized drives no longer find a responsive sound in the gathering silence of the disappearance of the social; the animal/human rubbing itself raw on the bars of oppressive customs and punctilious behaviour has been liberated by being abandoned as biogenetic road kill.

Bad conscience as the psychological parasite feeding on the host of human flesh is suddenly malnourished. It urgently requires another host, another carrier, another adventure and wilderness and homesickness. It takes a gamble. It leaps to the digital. It parasites the will to technology. It begins to feed on the flesh of the digital. It is a new theory of (digital) moral sentiments. It invests the digital nerve with the psychological motivation that it always desperately required: a techno-feeling of self-loathing, of resentment, being cheated. The digital nerve cannot exhaust its drives internally. It has broken with human flesh and thus can no longer use the now disappeared human as its alibi. It feels the weight of electronic lightness gathering in the interstices of its data nodes. It has no spirit of adventure because it is the end of seduction and the beginning of the regularity of calculation. It experiences no homesickness because the digital nerve is the electronic home that the will to will always sought, and now finally has found. It is not a torture chamber but a pleasure palace of fully realized technicity; not a cage but a network; not a place of chained animals, but a process of vectored flesh. But still the digital nerve is all drives and energy and force and empty quanta of power. If it cannot exhaust its drives internally, then it must project itself externally. It must harvest human flesh, nature, the sky, the earth, water, the moon, the planets, and, beyond, the galaxy itself. Stripped of its human carrier, the digital nerve is still a will to power. It feels cheated and aggrieved because it can never fully satisfy its transcendental drives. By definition, it is always a drive to (virtual) emptiness, to (electronic) space with no interiority of time, to an archaeology of speed without depth, to connectivity without the illusion of communication. In the end, a 'gamble' a 'transition,' the will to technology experiences the first intimations of the eternal recurrence of mythic fate. Now it is no longer 'man who is weary of man,' but technology which is weary of technology. The digital nerve feels itself a stranger in net time. The fibre network emits no satisfying response of digital reciprocity. The digital nerve considers suicide. But it cannot act in this direction since it is always already post-suicidal: that's the exterminatory quality of a technological drive to harvesting the human remainder. It grows melancholic and depressed, but it cannot stay this way for long because it is driven to fulfil its destiny of speed. A 'gamble' cannot admit that it is bipolar; a 'transition' cannot be a looking-back; an 'adventure' cannot be a gated community; a 'wilderness' cannot be soft domesticity. So the will to technology is forced to go on, to cross over, to be a leap, a transition, a dance over the abyss. It is weary of man, resentful of its human parasites; feels short-changed by its future; it is born not owing nature a debt. Like man before it, the digital nerve quickly reaches that point of 'monstrous consciousness' – great self-pity mixed with self-loathing – and, as such, the digital nerve becomes the spearhead of nihilism.

Nietzsche again, this time *On the Genealogy of [Digital] Morals*:

The man of [digital] ressentiment is neither upright nor naive nor honest and straightforward with himself. His soul squints, his spirit loves hiding places, secret paths and back doors, everything covert entices him as his world, his security, his refreshment; he understands how to keep silent, how not to forget, how to wait, how to be provisionally self-deprecating and humble ... slaves are clever.[17]

The Maggot Man

Not fear; rather that we no longer have anything left to fear in man; that maggot man is swarming in the foreground ...[18]

The 'maggot man' is the creative leader of virtual capital feeding on dead flesh. Himself a servomechanism of the machinery of capital, the maggot man is the last harvester of the human sensorium, transiting the human nervous system into the logistics of the digital nerve. Neither absolutely dead nor even virtually alive, the maggot man has a keen smell for the scent of living flesh buried under the electronic debris of media culture. Like a scavenger machine tuned to a fast recycle loop, the maggot man sniffs culture and music and street fashion and sex and drugs and the wasteland of human feelings, not a cultural stone is left unturned, no energy of the street is exempt from his attention, no rap rebel is too outlaw, no boredom suicide in a suburban shopping mall is too sacred. Like a cyber-dog on speed, the maggot man transforms living energy into electronic cairns of dead culture skin, and crawls inside. Now, it's the cinematic gaze and television eyes and fashion plug-ins and the streamed data net as the extended flesh of the maggot man. He wears many skins. He smears his face with many liquid masks. He is all an amplified nose rooting around in what's left of cyber-life. He is all a magnified ear listening through the noise spectrum to the sounds of desolation within. A specialized function who, as Nietzsche said in *Thus Spake Zarathustra*, 'knows too much about one thing and nothing about anything else,'[19] the maggot man, viewed from above, appears as a thin spindle of bio-wire connecting his over-specialized organ of perception to the trompe-l'oeil of his human flesh. He is always on over-drive because he has only got a future. He is possessed by the will to speed like a terminal disease. He's the possessed individual, and the possessive business flesh. The maggot man crawls over the skin of culture. He is a parasite/predator: the ascendant character type of contemporary culture.

Never acting alone, the maggot man is a swarm machine running in increasingly vicious hunting packs. We call them corporations in the marketplace, the state in public life, policing the homeless, hunting down the poor, sweeping the streets of squeegee kids, pepper-spraying human rights protestors in Quebec, Prague, Genoa, Washington, Calgary, Toronto, Montreal, Seattle, powering up the cyber-armies of the night, the human genome experimenters of the body – surveillance machines and credit machines and memory machines and forecasting machines and nano-machines and money machines.

A machinery of dead power led by what Nietzsche called 'blond beasts of prey.'

> I employ the word state: it is obvious what is meant – some pack of blond beasts of prey, a conqueror and master race which, organized for war and with the ability to organize, unhesitatingly lays its terrible claws upon a populace perhaps tremendously superior in numbers but still formless and nomad. [Or] one does not reckon with such natures; they come like fate, without reason, consideration, or pretext; they appear as lightning appears, too terrible, too different, too sudden, too convincing ...[20]

Under the sign of the digital nerve, some of Nietzsche's 'blond beasts' have left the state and gone into private (commercial) practice. The digital nerve wouldn't have it any other way. It needs its maggot men and blond beasts of prey. It rewards them richly. It sucks them dry. It makes them rich. It makes them powerful. It makes them arrogant. It makes them feel superior. It makes them feel transcendent. It makes them feel virtual, almost post-human, capable of doing anything, a better-seeing cyber-dog while morally blind. It gives them uniforms of the hunting pack. It needs a human face and a human scream – its 'conquerors organized for war and with the ability to organize.'[21] Wanting one thing only, to lay its cyber-claws 'upon a populace perhaps tremendously superior in numbers but still formless and nomad,'[22] the digital nerve is a war machine. It has smelled the plasma scent of solar flares and tasted the darkness of galactic space, and knows that it cannot come alive without the sustenance of human flesh. It is a data carnivore. It speaks through its ascetic priests: the 'maggot man' and the 'blond beasts' of prey.

This is a transitional period. We are transiting to the digital nerve. It is our goal, our value-direction, our ascetic ideal. The will to power splits: one side pushes ahead to virtuality; the other pole turns backward to feed on the flesh of human culture; one dreams of interfacing the digital nervous system; the other experiences the digital nerve as a war machine. A culture of technotopian eyes, with lips that speak abuse. Both opposing value-directions. But both value-directions necessary as part of the same logic of digitality. The 'maggot man' and 'blond beasts' of prey swarm into the foreground of the will to digital power. They feed on the open wounds of split culture, naming the whole thing the difference that animates. *Sacrificial power*: split between virtual management techniques and draconian policing of resisters. *Sacrificial minds*: split between the data mad and the memory dead. *Sacrificial politics*: split between the language of technocracy and abuse-value. *Sacrificial personalities*: ingratiating smiles for the powerful, rancour for the weak and the defenceless. Split knowledge: calculative but forgetfulness. A culture of nihilism. A culture of seduction. A culture of rap. A culture of brand flesh. Technology grown weary of itself meets man grown weary of himself: the result is the zombie-culture of the twenty-first century. But still, there are Nietzsche, Heidegger, and Marx murmuring in the background: speed capitalism, accomplished technicity, the will to power. And sometimes, just sometimes, there is Nietzsche's Zarathustra in conversation with the fire-dog:

> The earth has a skin; and this skin has diseases. One of these diseases is called 'Man.'[23]

And sometimes, just sometimes, there is Nietzsche in conversation with the future yet to be born, with virtual man as a self-overcoming – 'a dangerous going-across, a dangerous wayfaring, a dangerous looking-back, a dangerous shuddering and staying-still.'[24] This Nietzsche the futurist – has no illusions about the price to be paid for the bestiary:

> Even what you omit weaves at the web of mankind's future; even your nothing is a spider's web and a spider that lives on the future's blood.[25]

When the will is burned into man, then we are the disease that lives on the blood of the future. The fire-dog nods his assent, although maybe he was just looking for another bone along the way.

NIETZSCHE SCANS

Nietzsche Shareware

Expressing itself in the form of distributed theoretical code, Nietzsche's thought moves at great velocity within the circulatory systems of the social, strictly non-hierarchical, a flux of intense aphorisms not a dogma of fixed ideas, transgressing the closed boundaries of intellectual property rights, owned by no one, refreshed by everyone it touches, mutating and recombinant and intentionally non-linear, a secret writing for an over-exposed time of immense surveillance, a public code of rebellion operating within the language flows of a policed culture: a polemic, a criticism, an overcoming, a shareware storm that will not be denied.

An inveterate nomad, Nietzsche Shareware always seeks out the limit experience. A natural outlaw solitary and defiant in the linguistic universe of stable referents, Nietzsche Shareware sets in motion a seductive game in which the great markers of meaning – power, sex, consciousness, truth – are suddenly flipped inside out, ablating their previously hidden codes. Power unmasked as the will to virtuality. Sex as the insatiability of the death drive. Truth as one expression of the culture of boredom. Justice as sacrificial violence. The unconscious as an alibi for the loss of symbolic exchange. Intensity is its sign; the desire to overcome the despotic binaries its ruling passion. Mutation is its preferred way of becoming; recombinant logic the internal strategy by which Nietzsche's thought folds back on itself in an indefinite simulacrum of truth and anti-truth. Nietzsche Shareware is open-systems architecture: always accessible, brimming with viral energy, speaking in coded terms that are at once somehow interior to the deepest logic of the systems within which it operates and yet completely exterior in terms of its absolute unacceptability.

Like the circulating codes of shareware itself, Nietzsche is always hunted down: a double sign with the sex smell of a dangerous seduction and the hint of death of an impossible transgression. Consequently, for every policeman present at the utterance of Nietzschean aphorisms and for each Christian judge awaiting transgressors of the will to truth, the media galaxy of simulacra warms itself at the cold fire of Nietzsche's solitude. And why? Because we are all Nietzsche Shareware now: all circulating within the cultural codes of the 'will to power,' and the 'maggot man' and the 'last man' and the 'will to will' and the 'world as a hospital room' and *Ecce Homo* and 'ressentiment' and the 'ascetic priest' and the delirious poetry of *Thus Spake Zarathustra*. Ironically, Nietzsche Shareware is the purest expression possible of the digital future, both in terms of its 'completed nihilism' and those other shareware stars dancing over the digital abyss.

Nietzsche at the Montreal Pool Room

3:00 a.m on an early Sunday morning, and I'm parking my flesh on a silver stool in the back-end of the Montreal pool room. Not that there's a pool table, there hasn't been for years. Just a slabby wide open space with tin silver counters running along the walls, video poker machines, glaring fluorescent lights, and a grimy order counter jammed by the front door. The kind of lunch counter where you flick in your body off Boulevard St Laurent, smell a hundred years of fried grease, burnt onions, sour relish, and just know that you've got to have some. So, I've got a copy of the *Journal de Montreal* for street-time reading pleasure about all the crime and the deals and the dead and the stars and the Rock Machine biker wars and the politics gone wrong and sometimes gone right that make up Montreal, pucker your lips into the just-so-incorrect east-end tongue, and order a couple of chien chauds 'all dressed and make that double onion, please,' one big bag of ketchup-drowned frites, and some hard-luck coffee. When you get your food, which is instantly since this a cash on delivery and fuck you buddy kind of place, you take your mouth and your eyes and your cooled-down nerves to that empty stool at the back, vector a chien chaud, taste those perfect fries, and settle back for a good data-read of the *Journal de Montreal*.

Except you never get that far because Robo-Dean appears at the door, spots you right away, and hustles over with breaking news on Nietzsche. He's got great street-smart body armour: shaved head, dark shades for better X-ray vision, long black leather coat, and a tattoo of SPIT stitched across his forehead. But that's all beside the point because he's in a hyper-trance mood: no sleep, just the right mix of happy-time drugs to open up the wonder pores, and a multi-task read-through copy of Nietzsche's *On the Genealogy of Morals* in his hand.

Maybe a wild-eyed vision of Nietzsche in his last Turin days, the time when he just finally stopped writing and went home to the silence of his inner self, doing one final write on his body and mind and nerves and soul in ruins. Sort of like a virus that sometimes goes underground for a time, and just sometimes also slips out of the air to take possession of another wandering minstrel of the night. Like Robo-Dean, who doesn't even wait for hailing distance, but shouts across the Pool Room.

'Hey Cloner! Wake up, it's Nietzsche time. Did you ever read that passage in the *Genealogy*, the one where Nietzsche talks about the pleasures of cruelty? About how only pain hurts, and so the ascetic priests put burrs in our flesh, little memory-reminders, to keep us all in line?'

Now I wasn't none too happy to get dream-jumped on my fries and hot dog and *Journal*, but I'm a sucker for Nietzsche, and if he's decided to pay me a visit on this early morning of the Lord in the Montreal Pool Room and in the likely person of Robo-Dean, then what the hell, let's get to it, and see what visions crazy, sad, mad, and maybe just keen-eyed wise this visitation of Nietzsche is all about.

Because I know this. It would be just like Nietzsche to flesh-morph in the earthly form of Robo-Dean, saunter through the door of the Montreal Pool Room, and lay down some new aphorism tracks for decade zero. And it's sort of cool. Just when you think you've left Nietzsche long behind and you're settling into your own groove of maybe settling for less for but settling none the less, he suddenly whomps up in the middle of a night-time street scene, cackling and groaning and whining and bitching. And you don't necessarily want to listen to him, you may not even want to read him anymore, but he's got your cell phone number, and you know you're netted in his spider's web. Nietzsche even predicted it in advance. He once said: 'Now that you've read me, the problem is to get rid of me.'

So, with just a little murmur of what-are-you-doing-in-my-face discontent, I jump Robo-Dean with some fast probes.

'Why not? Nietzsche is the medium. Not just little burrs, but now digital burrs, little electronic trodes cut into the flesh.'

Robo-Dean flashes a jack-happy smile. I've made a connect.

Channelling Nietzsche

And I was right because Robo-Dean sits right down on the next stool, takes some of my fries, and tells me straight-out that he's got a story to tell. Something about channelling Nietzsche. But first, he looks up at the mega-sunshine fluorescents and says: 'It's too bright here for Nietzsche. Let's go to Nausea.'

Which was fine with me because Nausea is a Nietzsche-like bar on St Catherine. Definitely not a cyber-café, it's where all the prostitutes and transsexuals and drug dealers and pimps and philosophy students and slumming hackers from Softimage or maybe even Behavior up the cyber-way go to get one last fix of night-time spirits to see their way through to the morning light. A shot of scotch in one hand and a beer chaser in the other, Robo-Dean rocks on his heels and in that rabid voice that just jackhammers away at your nervous system with no apparent breathing holes, he looks me in the eye and asks: 'Have you every channelled anyone?'

When I admit right off that I haven't, Robo-Dean declares, 'Well, I have. Last night I channelled Nietzsche, and he's got a message for you, actually a disk.'

Robo-Dean hands me the disk, and I flip it into a vector portal at Nausea, see a cute 4-D multiplex image of Nietzsche as he might have looked in his love affair with Lou Salomé days, and read what looks at first like an introduction to a new text titled *The Digital Nerve*.

Digital reality as the final story of Christianity. Clonal engineering, synthetic chromosomes, burning new genetic codes into the flesh: what are these but last signs of the viciously naive will? Exhausted with life, tired of dragging flesh on its death-march to the grave, the will fatigued, in lassitude, unable to believe in its own myth, unwilling either to go forward or to close time's door, the will declines to will, the will abandoned to the will-not-to-will. Digital reality not as simulation, but as an alternative reality, an artificially engineered reality of clonal flesh and synthetic nerves and android chromosomes. Two wills, two bodies for the millennium, divided and at war. The tortured body of the last remains of will-less Christian flesh, and the cynical will of the digital nerve. Has the will become a clone of itself? Which will triumph? The body as a vivisection-machine? Or the digital nerve as a successor species to a humanity taking cynical pleasure in willing its own disappearance? How long can the body tolerate its radical separation into two species-forms? And what beast of the virtual will arise from the graveyard of this meeting of great pity and great nausea? On the Genealogy of Digital Morals *as the tombstone of Christianity in its final resurrection-effect as the sign of the virtual beast. The epochal dreams of digital reality are not so far away from the deserts of North Africa in the fourth century, that moment when St Augustine triumphantly severed flesh from spirit, beginning the search for our successor species, first in the torture chambers of absolute religion, then in the war zones of absolute ideology, and finally in the futurist algorithms of absolute technotopia. But I anticipate Camus: the union of absolute justice and absolute reason equals murder in the name of freedom. The question remains: Is digital reality the final act of species murder, the (human) blood sacrifice necessary to inaugurate the reign of the post-human? But that would be a question of myth, and mythical thought, most of all, is denied by the feverish and calculated positivism of the new codes. Nihilism today speaks in the algorithmic tongue of the digital nerve.*

The Digital Nerve? Life as an edge between Nausea and the Montreal Pool Room.

NIETZSCHE MICROSOFT

Consider the will to power of Microsoft, that most Nietzschean of digital corporations. Here, the will to power is transformed into a pure axiomatic of digitality: a will to virtuality that is almost metaphysical in its insistence that the mathematical structure of the Microsoft 'method' provide both the condition of possibility and justification for the triumph of the digital nerve. While Microsoft can be analysed economically in terms of predatory marketing practices, technically in terms of the quality of its software products, and culturally in terms of its contribution to the unfolding story of technology and the American mind, it is only by rubbing Microsoft's vision of the digital nerve against Nietzsche's *The Will to Power* that the essence of Microsoft as the leading contemporary expression the will to will discloses itself. Microsoft is Nietzsche.net.

The Will to (Digital) Power: Technology @ the Speed of Business

The will to power can manifest itself only against resistances; therefore it seeks that which resists it – this is the primeval tendency of the proto-plasm when it extends pseudopodia and feels about. Appropriation and assimilation are above all a desire to overwhelm, a forming, shaping and reshaping, until at length that which has been overwhelmed has entirely gone over into the power domain of the aggressor and increased the same.[26]

If Microsoft is the essence of the will to digital power, then studying Microsoft reveals the deep archaeology of information technology, what Nietzsche.com comes to mean in the streets of data marketing and softculture and distributed knowledge. In the same way that Nietzsche focused his meditations on the metaphysics of Christianity as the breeding-ground of ressentiment, so too understanding Microsoft, examining, that is, the will to digital power as post-human destiny, can best be accomplished by a careful reading of its main metaphysical text, Bill Gates's *Business @ the Speed of Thought*. This is theology for the digital way – a euphoric and uncritical vision of the merger of flesh and machine in the wireless nodes of the 'digital nerve.' Precisely because this text is uncritical of its essential assumptions and in that spirit of uncriticality perfectly transparent about what needs re-engineering as a requirement for the triumph of the digital nerve, this book can be read with a sense of dread and fascination: dread because it announces the forced integration of every facet of human experience into the 'digital nervous system'; and fascination because, as a self-confident statement of the virtual class at the beginning of its post-human reign, the text is ideologically transparent – a direct read-out of the global brain of the will to digital power. It is a Nietzschean feeding-frenzy. If Nietzsche were alive today, he would be a writer of XML programming codes, listening to DJ Spooky with one cyber-ear, with a cyber-eye attuned to Microsoft's newest Internet strategy – Micro.Net with its sudden extension of digital ganglia for harvesting the whole world wide web, the proof of which is the disingenuous comment by Steve Ballmer (president of Microsoft): 'Are we a company that is going to suck up everything into itself?'[27] Microsoft as Nietzsche's 'pack of blond beasts of prey, a conqueror and master race which, organized for war and with the will to organize, unhesitatingly lays its terrible claws upon the [digital] populace perhaps tremendously superior in numbers but still formless and nomad'[28]

Consequently, more than another tech hype book in the data vector, Bill Gates's *Business @ the Speed of Thought* is simultaneously a manifesto for the triumph of digital business as the dominant ideology at the cusp of the twenty-first century and a dynamic, well-theorized, and ultimately chilling description of the business strategies involved in 'using a digital nervous system.' Not so much 'business @ the speed of thought' as technology @ the speed of business. Here, the creative possibility that was the digital future is effectively shut down in favour of a closed business culture that takes electronic culture and hijacks it as a way of powering up digital capitalism.

If Gates's conception of digital business were simply a continuation, even an intensification, of the rhetoric of traditional capitalism, that would doom his perspective, and his future business strategies, to the normal ebb and flow of the cycles of capitalism. However, what makes Gates's perspective such a radical rupture in the rhetoric of competitive, although always monopolistic, multinational, capitalism is that Gates is both the author of a biological model of digital business and an astute business theoretician of the specific strategies necessary for booting up the 'digital nervous system' as the operational language of, at first, business and then later those other 'special enterprises' – education, medicine, government, warfare. What is disclosed in this book is nothing less than a general political philosophy – a digital Walden 2 – with Gates as B.F. Skinner's overseeing manager installing the 'digital nervous system' in business, in education, in human flesh, in public policy, in the biogenetic body, in cyberwar. This is a book not so much about business as about the nature of power – cyber-power – in the 'digital information flows' that code electronic culture. Much like the ruse of the Trojan horse in Homer's Iliad, it may well turn out that Gates's lasting importance lies in both installing, and then making come alive, a biological model of technology in the seductive form of the 'gift' of digital business. Business @ the Speed of Thought, then, as a post-human model of business for a post-business conception of technology. True to the implied destiny of his name, Gates is an astral gateway to a digital future of wired flesh.

Ramping Up the 'Digital Nervous System'

Gates is explicit about the biological basis of digital business. Consider the following:

> An organization's nervous system has parallels with our human nervous system. Every business, regardless of industry, has 'autonomic' systems, the operational processes that just have to go on if a company is to survive...What has been missing are links between information that resemble the interconnected neurons in the brain.[29]

> You know you have built an excellent digital nervous system when information flows through your organization as quickly and naturally as

thought in a human being and when you can use technology to marshal and coordinate teams of people as quickly as you can focus an individual on an issue. It's business at the speed of thought.[30]

What systems theorists such as David Easton and Norbert Weiner could only postulate in the 1960s, Gates puts into actual political practice in the twenty-first century. In Gates's cyber-world, the 'feedback loops' of general systems theory merge with the dynamic logic of biogenetics to create a post-human vision of digital business. Here, there are no human beings, only 'inflection curves'; no digital dirt, only 'interconnected neurons in the brain'; no accidents, only 'autonomic systems'; no history, only 'data mining'; no human vision, only 'pivoting the data from every angle.'

An analytically abstract, fast circulating, highly coded feedback loop of 'good digital information flows' and 'good analytical tools,' Gates's model of post-human business is the key interface by which human flesh will migrate to the machine in the digital future. Once fully operational, the digital nervous system can be quickly installed in every form of organization. Microsoft is only apparently about products. In reality, it's about a certain procedure, a certain form of cybernetic organization that, once installed, patterns digital information flows across the nervous systems of all the key institutions of contemporary life. Microsoft, then, as not so much a 'global brain' but as a downloadable, ready to install, virtual memory: a cyber-Panopticon plugged into the flesh circuits of human subjectivity. The virtual architecture of the future, the digital nervous system boots up IT Being into living existence. Cybernetics finally comes alive, or as Gates likes to say: 'Information flow is your lifeblood.'

The Ideology of Information Technology

Information technology and business are becoming inextricably interwoven. I don't think anybody can talk meaningfully about one without talking about the other.[31]

There is nothing more relentlessly ideological than the apparently anti-ideological rhetoric of information technology. Here, the question of ideology breaks with the received interpretations of ideology as 'false consciousness,' as 'spectacle,' or as the historical spearhead of opposing class interests, sinking instead into the deepest tissues of everyday reality as the transparent, technical supporting infrastructure of digital life. The ideology of information technology, then, as a (business) gene machine sequencing human experience into the working data flows of a cybernetic system running flat-out on automatic. It's what Gates describes as 'manag[ing] with the force of facts.'

If, in practice, ideology is understood as a rhetoric machine that projects into power an underlying vested interest by presenting as the general will the specific interests of a particular will, then Gates's 'management with the force of facts' is the precise rhetoric machine that spearheads the emergent class interests of the virtual class – the dynamic, although inherently unstable, coalition of 'knowledge workers' and digital capitalists dominating the e-commerce and e-government and e-medicine and e-education and e-war of the twenty-first century. In *Business @ the Speed of Thought*, the general will merges with the digital will, and the digital will is reduced to the technical, cybernetic procedures involved in booting up the digital nervous system, namely standardization (of digital information flows), surveillance (of knowledge workers and digital customers), subordination (of human intelligence to digital intelligence, of human flesh to machine flesh), and solicitation (of particular wills by the general will of digital reality in the name of greater cyber-communication, better digital knowledge, 'raising your corporate IQ,' 'empowering people,' 'creating connected learning communities,' 'preparing for the digital future'). In the Microsoft rhetoric machine, ideology always interfaces digital subjectivity. Consequently, the four leading war tactics of Microsoft digitality – standardization, surveillance, subordination, and solicitation – are the dynamic expressions of the hegemonic ideology of the virtual class, thinly disguised as forms of digital technicity. That's why *Business @ the Speed of Thought* could be a bestseller: not only the charismatic lure of Gates as the 'world's richest person,' and not simply the now familiar recipes for 'friction free capitalism,' but also because this book is a rhetoric machine, laying down the key codes for translating into political, which is to say business, practice the particular interests of a global digital elite.

In the fully realized world of technocracy, ideology is a gene machine.

'To Think, Act, React and Adapt'

Early in his book, Gates confesses the intellectual origins of his business philosophy: a mixture of Alfred P. Sloan's *My Years with General*

Motors and Michael Dertouzos's *What Will Be: How the New World of Information Will Change Our Lives*. In other words, the organizational genius of General Motors in 'standardizing' large-scale business practices and the digital vision of 'augmented reality' authored by MIT's Laboratory for Computer Science. A perfect fusion, then, of information technology and business as the basis of the digital nervous system.

Gates is so drawn to Sloan's business philosophy ('It's inspiring to see in Sloan's account of his career how positive, rational, information-focussed leadership can lead to extraordinary success')[32] and his vision of the digital future is so repetitive of Michael Dertouzos's rhetoric of 'augmented [virtual] reality' because Gates's peculiar skill lies in combining a rhetorical veneer of technological utopianism with the business reality of a 'by the numbers' large-volume, networked distributor of an off-the-shelf 'digital nervous system' for all the 'special enterprises' of global culture, from corporations to governments, education, medicine, insurance, banking, and the (cyber-)military.

In the nineteenth century, Hegel might have diagnosed in advance Gates's Microsofting of the world when he theorized the charismatic importance of 'world-historical individuals' – subjects who sum up in their personalities the ruling geist of the times. Writing in the turbulent aftermath of the Battle of Austerlitz, Hegel had the victorious generalship of Napoleon in mind, but given his philosophical commitment to the coming-to-be of the 'universal rational state' surely he would not have been disappointed by the appearance on the (digital) historical scene of Bill Gates, this 'world-historical individual' whose technically obsessive, hyper-pragmatic, relentlessly acquisitive vision of installing a 'digital nervous system' as the networked ganglia of the global digital body actualizes with a vastly larger historical sweep and certainly with greater messianic enthusiasm the idea of a universal rational state than any of the particular military victories of Napoleonic campaigns. While force of circumstance and limits of technological development limited Napoleon to the European theatre as his stage of historical action, Gates's vision of the digital nervous system operates on a planetary scale, downloaded here in Microsoft India and Microsoft France, uplinked there in Microsoft Office and Microsoft Exchange, streamed data flows in Microsoft Sales, 'data drilling' and 'data mining' for business opportunities 'pivoted' for better digital vision – the 'universal rational state' ground down to finite particles of data, revealed with brilliant 'granularity of detail.' If Hegel were alive, he would immediately take to pen again to write another *Philosophy of*

[Digital] History with Bill Gates as the historical subject who best expresses the ruling spirit of the times.

That Gates himself is hostile to 'philosophical discussions of whether this is a Sun Belt or Rust Belt City,' preferring to 'drill data' deep down, to 'pivot' data from impossible angles so as to tease out the 'mathematical story' of markets unconquered not yet told, does him no disservice. In exactly the same way that Hegel said that history is always dialectical, always an active opposition between the antithetical poles of immanence and transcendence, so too Gates's vision of the digital nervous system reconciles the contemporary dialectics of business history, unifying, for example, the grim organizational immanence of Sloan's genius for 'standardization' so aptly represented by the heavy-image weight of General Motors with the spuriously transcendent vision of 'augmented digital reality' represented with such utopian enthusiasm by Dertouzos's MIT Laboratory for Computer Science. Indeed, if Gates in *Business @ the Speed of Thought* can speak so decisively of 'action numbers' and free-information flows and ceaseless network surveillance and 'punctuated chaos,' if, that is, he writes as if in an animated state of always falling forward into decisive business decisions, that's also because he is himself the first and perhaps best of all the 'action figures' or maybe a digital GI Joe – a leading business personality who has thought so deeply and so well of the business applications of speed data and speed technology that he has become a 'punctuated chaos' or, in his own terms, a personality of 'constant upheaval marked by brief respites.'

And this is exactly as it should be because Gates has already claimed that 'information work is thinking work.'

> When thinking and collaboration are significantly assisted by computer technology, you have a digital nervous system. It consists of the advanced digital processes that knowledge workers use to make better decisions. To think, act, react and adapt. Dertouzos says that the future 'Information Marketplace' will entail a 'great deal of customized software and intricately dovetailed combinations of human and machine procedures' – an excellent description of a digital nervous system.[33]

For Dertouzos, human flesh will assist the cybernetic development of an augmented digital future. Human flesh will become the machine flesh of the digital future. Or, in Bill Gates's case, the digital nervous system that he so eloquently espouses can be articulated with such main business force and compelling managerial confidence because as the leading exponent, and beneficiary, of the information marketplace, Gates is the first 'intricately dovetailed combination of human and machine procedures,' the first living digital nervous system. Gates is already internal to that which he theorizes. He thinks the digital universe from the inside in terms of its unfolding matrices of business mathematics, and he can actually feel the architecture of a 'three-tier computer interface' system. As the extended mind of Microsoft he spins its digital wheels while 'waiting for the rest of the world to catch up.' Gates is the digital operating system that he thought he was only licensing.

An American (Digital) Revolutionary

A decidedly local thinker with global ambitions, Gates is only the latest representative of the spirit of rationalism that is the American business creed. While contemporary media culture in the United States assigns the Civil War an 'augmented' historical importance as the determining moment in American political history, Gates's book is a sharp reminder that the political DNA of America was genetically mixed in the Revolutionary War, and that the same ultimately bourgeois revolutionary spirit that broke with the British empire on the grounds of ' life, liberty and the pursuit of happiness' fostered a spirit, at first national and then world-conquering, of unfettered individualism and market competition. A single-minded spirit of action as being, trading as salvation, doing as morality, an American spirit that has always found its most enthusiastic representatives in the business class. What makes the American spirit a 'world-historical idea' and what makes Gates a 'world-historical subject' is that the American Revolution, while fought on the political grounds of constitutional and economic independence from European colonialism, installed in the New World the very form of unconstrained and unfettered human subjectivity – ''life, liberty and the pursuit of happiness' – and the very expression of pragmatic and positivistic and wilful human consciousness that was necessary for the remaking of America as a radical experiment in technology. In this (American) country and in this (technological) revolution, the spirit of European rationalism as the intellectual backbone of what would come to be called the empire of technology could only fitfully and episodically be expressed, constrained here by agrarian class interests, fettered there by the political status quo, but not so in the American Revolution. In the latter, the essential spirit of technology, the spirit of endlessly remaking and reinventing the self and the world, was what was at stake, and what was ultimately won in the American revolutionary wars was nothing less than a Declaration of (Technological) Independence. In the 'new morning' of America at Lexington and Concord, it was also the self as technique – American individualism as an open-ended process of struggling for life, liberty, and the pursuit of happiness – that was born as an animating spirit of world history.

It is in this sense that *Business @ the Speed of Thought* is, above all, *political theory*: a projection into world history of the animating idea of the American Revolution, although modified to conform to the spirit of the digital future with, for example, the development of a 'web lifestyle' substituting for life, 'augmented [digital] freedom' for liberty, and data commerce as the electronic equivalent of the pursuit of happiness. Which is to say that the fundamental historical importance of Gates's vision of the digital world is that it simultaneously recovers and reinvents the American revolutionary creed for the digital future. Long before the appearance of the digital world, what made American political culture distinctly modern and radically different from its European genealogy is that American identity was always coded, always booted into existence by a trinitarian code. What was really invented at the Continental Congress was the original American code as the dialectic of life (immanence) and liberty (transcendence) with the 'pursuit of happiness' as the mediation between the two. This trinitarian formulation was the essential spirit of the American will. Dynamic, restless, always given witness by action, regenerated by violence, strengthened by opposition, steeled by adversity, enthusiastically pioneering, going forward, future-bound, indifferent to the passing contents of its particular manifestations, this distinct expression of the language of the will flows directly from the American Revolution to *Business @ the Speed of Thought*. And why? Because Gates's vision of digitality is extreme will: the American trinitarian code stripped of its covering rhetoric and its historical baggage, digitized and genetically booted, ready to be system-installed in the twenty-first-century American mind. Here, the language of the will, at first anti-colonial, then pro-bourgeois, and finally fully digital, abandons its temporary refuge in the body of the enterprising American individual, breaks with its strictly capitalist determination, taking up a new residence as the animating spirit of the digital nervous system.

Neither purely immanent nor solely transcendental, the digital will is both simultaneously. It is the virtual will, moving first at the speed of business and only later in its future appearance as unfettered technicity at the speed of light.

'Three-Tier Computing'

A computing architecture in which software systems are structured into three networked tiers or layers: the client or presentation layer, the business-logic layer, and the data layer. PCs usually provide the presentation layer, PC servers the middle tier, or business-logic layer, coordinating relations between the user (client) and the back-end tier. The data tier often includes a variety of PC and non-PC systems.[34]

The business model can be hegemonic today as the spearhead of digital reality because in the particular expressions of acquisitive business interests are to be found, in however imperfect a form, the general manifestations of the digital will.[35]

That's the specific importance, for example, of 'three-tier computing.' Not so much a vision of the future of digital business as a specific model of the digital will which has already been installed at the nerve centres of networked society. In the smooth, seamless circulatory flows of 'three-tier computing,' the system as a whole is front-ended by the disappearance of the human (into cyber-clients), mediated by business-logic, and back-ended by waiting data warehouses for remaindering virtual memory. A perfect cybernetic system, a smooth, although tentative and yet incomplete, expression of the operating logic of the digital will. In the cyber-world, the digital will no longer expresses itself directly in the language of warfare or philosophy, but in the practical language of business, this seemingly non-violent language of studied superficiality and dynamic pragmatism.

Consequently, studying the business model in general, and Gates's description of the digital nervous system in particular, is an exercise in political theory. The language of digital business is a precise diagnostic of the power points of the emergent digital future. Consider, for example, Gates's favourite buzzwords – 'disintermediation' or 'friction-free capitalism.' On the surface, these terms point to a digital (business) future in which the middleman is removed from a transaction, replaced by direct meetings between producers and consumers transacting on the Internet. A 'friction-free capitalism' of e-buys and e-sales, with added value from the absence of the parasitical middle.

Except, of course, there is always a hidden third party – the computer interface – and it is already under two forms of monopoly control: first, Gates's licensing of the digital operating system; and secondly, Gates's creation of an extensive virtual network of on-line sales. Friction-free capitalism perhaps, but certainly not fiction-free. Or, as James Glick recently wrote in the *New York Times*: 'Soon Microsoft will collect a charge for every airline ticket you buy, also every credit card purchase, picture you download, web site you visit.'

In the Microsoft model of the digital future, we have been transformed in advance into 'dumb clients' and 'dumb servers,' 'data mined,' 'data mart-ed,' and 'data warehoused,' 'feed-back looped,' 'groupwared,' 'supply chained,' and 'electronic data interchanged' into 'good digital flows' for better 'horizontal integration' and 'executive information systems' – 'just in time' portals for 'plug and play' in the 'paperless office' of digital reality.

The Politics of Digital Excess

Gates is correct in one important respect. *Business @ the Speed of Thought* provides a way of 'rid[ing] the inflection curve' to a detailed understanding of the present and future computer architecture of the digital future. Gates admonishes us to 'adopt the web lifestyle,' to 'build digital processes on standards,' to 'develop processes that empower people' by 'know[ing] your numbers' and 'rais[ing] your corporate IQ' and 'treating IT as a strategic resource.' Because he is the spearhead of digital business, he is also a brilliant guide to its deployment of a global hegemonic ideology, in self-congratulatory chapters hyping 'Managing Knowledge to Improve Strategic Thought,' 'Bring Insight to Business Operations,' and 'Expect the Unexpected.' Not just a detailed analysis of the political methods by which business reduces Information Technology to its own acquisitive ends, *Business @ the Speed of Thought* also provides a privileged insight into the future colonization by digital business into four strategic 'special enterprises': health care – 'No Health Care System Is an Island'; public policy – 'Take Government to the People'; warfare – 'When Reflex Is a Matter of Life and Death'; and education – 'Create Connected Learning Communities.' Dispensing with the breathless missionary enthusiasm of technotopians, Gates's prescriptions for the business takeover of these four 'special enterprises' under the sign of the 'digital revolution' provide us with an early read-out of the methods (disintermediation), propaganda (digital necessitarianism), justificatory assumptions (speed, efficiency, and simultaneity), and ends (profitability) by which the key institutions of public life will be compressed into the 'three-tier architecture' of 'friction-free capitalism.' Precisely because Microsoft has the technical ability, economic means, and political will to impose the reality of *Business @ the Speed of Thought*, this book is, above all, a futurist manifesto with a difference. It predicts a future that it has the digital means to create.

Nietzsche's Strategic Managerial Analysis

If the government now can challenge Microsoft so relentlessly, perhaps that's because it is a challenge to its own power. A corporate noosphere that serves no nation because it is transnational, that advances no particular perspective because its operating system is a perspectival simulacrum, that is restrained by no specific interests because its primary interest lies in imposing its own 'interpretation' of the digital nerve on cyberculture, Microsoft is the will to (soft) power, and Bill Gates the 'ascetic priest' of digital reality. As in religious festivals of old where the function of the ascetic priest was always the twofold one of inspiring fealty among the congregation while keeping open the wounds of old resentments, Gates as the ascetic priest of the digital world is Nietzsche's übermensch, the 'overman' whose ascetic task lies in establishing the value-direction of the softwaring of human flesh. And if the government considers splitting Microsoft to be a fatal blow to Gates's ascetic leadership, then Nietzsche always knew better.

> If this incorporation is not successful, then the form falls to pieces; and the duality appears as a consequence of the will to power; in order not to let go what has been conquered, the will to power divides itself into two wills (in some cases without completely surrendering the connection between its two parts).
>
> 'Hunger' is only a narrower adaptation after the basic drive for power has won a spiritual form.[36]

'The will to power divides itself into two wills'? That's Microsoft divided into two companies – one an operating system, the other software products – the better to grow itself on the skin of the digital world as it searches about, seeking nourishment, finding resistances, assimilating, appropriating, overwhelming, extending (digital) pseudopodia, information-gathering, data-mining, conquering anew 'until at length that which has been overwhelmed has entirely gone over into the power domain of the aggressor and increased the same.' Microsoft is the essential contemporary expression of the *The Will to Power*, and Nietzsche is its leading strategic managerial analyst and digital anti-(soft)christ.

6 Streamed Capitalism: Marx on the New Capitalist Axiomatic

Streamed Capitalism

What is the significance of Marx in the age of globalization, at a time when the IMF can speak of installing 'automatic stabililizers' on the nervous systems of national political economies, when 'enterprise software' in the form of automated business-customer relations and business-to-business digital transactions has become the dominant code of streamed capitalism, when capitalism has suddenly and irreversibly speeded up beyond necessary production, beyond definite consumption, achieving for the first time in financial history that long-sought state of economic (digital) equilibrium: a zero-time circulation of value in a new economy typified by the circulation of pure capital? What is the significance of Marx in a culture in which technology and capitalism are effectively electronically streamed together, producing a hybrid version of capitalism for the twenty-first century? *Recombinant capitalism*, neither pure technology nor uncontaminated commodification.

Ideologically conceived in the era of digital utopianism, streamed capitalism consolidates its power in the contemporary era of economic recession. In the name of global competition, labour as a factor of economic production is politically disciplined in order to enhance the international fungibility of the electronic marketplace. For purposes of 'increased efficiency,' the previously fractured elements of competitive 'national' capitalism are decisively transformed into multinational archipelagos of networked capital, in key areas extending from energy to telecommunication and biotechnology. To maintain political stability in a time of viral terrorism, the state is presented with the double task of administering austerity programs under the auspices of the IMF and foiling the popular forces of the anti-globalization movement by designating the meeting sites of world financial leaders as 'military exclusion zones.' Having liberated itself from the inertial drag of economic territorialization and embodied labour, streamed capitalism makes its first triumphant appearance in world history in the virtual form of a new economic despotism. Paralleling the transformation of global military power into a new order of 'space war,' global capitalist powers now prepare for a new regime of economic competition under the war-sign of the spatialization of capital.

Ironically, it may be that streamed capitalism allows Marx's writings, notably the volumes of *Capital* and the *Grundrisse*, to be understood in their radical intensity for the first time.[1] Understanding Marx, then, not as a surpassed ideology given the radical structural changes precipitated by streamed capitalism, not as a labour theory of value in a period typified by the triumph of the (digital) knowledge theory of value, not as an analysis of capitalism as a production machine, but understanding Marx as an *epic thinker* with the rare ability to comprehend the historical unfolding of capitalism from its violent birth in the bourgeois struggle against feudalism to its triumphant maturity in the present age, that age prophesied long ago by Marx as the attainment of a form of advanced capitalism wherein 'value valorizes itself';[2] where the fetishism of the commodity-form gives way to the fetishism of money, and where the time of (capital) circulation approaches degree zero – capitalism as sub-infinity.

Understood projectively, Marx's epochal theorization of the history of capitalism finally realizes itself in the hyper-reality of virtual capitalism. What Marx could only predict as the inevitable movement of capitalism beyond production to the cycle of consumption, and thereupon beyond consumption to capitalism as a pure vector of circulation, now finds material historical reality in the mode of digital production. In virtual capitalism, capitalism as the dominant historical movement of modern and postmodern times sheds its guise as a mode of economic production, breaks forever with its epistemological preconditions in a logic of dialectics that situated use-value and exchange-value as necessary antinomies, finally announces triumphantly that all is now 'alternation' – labour and capital most of all – stripping away the outmoded skin of industrial and service capitalism to make its historical appearance in the twenty-first century as pure virtuality.

Marx as a Dark Futurist

Writing a century and half ago, Marx had the particular theoretical

genius to see right through the opening moments of industrial capitalism to its closing moments in the form of virtual capital moving at the
speed of circulation. Ironically, as a thinker who was historically fixed
in the immediate vicissitudes of his political experience, as a theorist
whose analysis was a war machine against competing political theories
and movements, Marx was always doomed, sometimes against his
own intellectual and political interests, to be a futurist. Always a theorist more than a polemicist, an existential writer more than a political
dogmatist, Marx in his thought always lived in a doubled space, between past and future. Not stopping to speak only of the 'double
moment'[3] of capitalism – its necessary logic of reversibility that was the
animating energy of *Capital* – Marx's writing was itself necessarily
doubled. Certainly, the continuing political appeal of Marx has to do
with his really existent historical analysis of the genealogy of capitalism
and its labour resistances. However, the real wisdom of Marx has to do
with his dark futurism, with his brilliant aperçus concerning a coming
stage of capitalism wherein value valorizes itself, commodity-fetishism
slipstreams into money-fetishism, and the circulation of commodities
gives way to the commodification of circulation.

This strange form of capitalism, this speed capitalism moving as a
lightning vector of circulation, can be recognized today as the first sign
of virtual capitalism. Ironically, only in the twenty-first century can the
prophecies and the history of Marx's *Capital* be actually realized in their
immediate political and theoretical form. In an ironic reprise of Foucault's
commentary on Hegel, just when we thought we had finally slipped
the trail of Marx, there waiting for us in the dawn of the new century is
the figure of Marx, a philosopher of the virtual future. For a rereading
of *Capital*, then, not as icon or as ideology, but as a not yet realized
political theory only now awakening to its philosophical possibility as a
futurist manifesto waiting in the shadowlands of thought as its long-
awaited nemesis, capitalism, moving at the speed of light, drops its
disguise as a model of production, becoming the historical possibility
that was always its hidden sign of seduction.

Capitalism, then, *as* the name given today to the movement of the
will to technicity. Capitalism can be so dominant today because it is the
historically realized form of *pure will*: pulsating, self-determining, breaking with all the (modernist) referents, abandoning any pretensions of
coming out of circulation to save the appearances of the models of
production or consumption, radically anti-dialectical, refusing commodity-fetishism in favour of the fetishism of signs, substituting the
knowledge-theory of value for a now objectively residual labour theory

of value, finally free to take its place as the centre of the historical nebula as a 'relation, not a thing.'[4] And Marx? He was the thinker who, understanding immediately the dark destiny of capitalism as pure will, devoted his political theory to a desperate search for an objective contradiction to virtual capitalism. As for ourselves, it is our fate to live in the midst of the dark destiny that Marx could only postulate, to live the future of fully realized capital as pure will, sometimes as emancipation, at other times as a digital carceral, but at all times with the understanding that Marx himself first realized but always sought to repress, namely that streamed capitalism may not only be the destiny of the west as its historical incubator, but the key cultural pattern of globalization in the new century.

Beyond his time and probably beyond his own (subjective) understanding, written like a theory machine running on automatic (de)stabilizers, Marx's epic traced the history of capital, first and foremost, as a 'relation, not a thing.' This was a thinker courageous enough to ask: *What if capitalism never came out of circulation?* What if capitalism implodes into a circuit of circulation that spirals inward on itself, enfolding and co-relational with itself, moving with such main vector force that capitalism eliminates all the signs of (industrial) capital with its crushing density? Consequently, two epochal hypotheses about virtual capitalism as pure circulation: *first*, the future of virtual capital as running on empty – no indefinite production, no necessary consumption, no romanticism of use-value, no exchange-value, no dialectic, only a cycle of virtual exchanges moving at the speed of circulation. *Or just the reverse*: hyper-capitalism as an explosion of production and a feast of consumption, a period of alternating excess and recession, fetishes everywhere and always, alternation of all the signs with no stability because the speed of capitalism has achieved the velocity of economic vertigo. In either case, it's Marx as epic theorist who first forecasts the realized history of the (capitalist) future as taking the form of capitalism as a sign of virtuality.

With Marx, the fetishism of circulation and the hyper-realization of the market *is* the dark future of *Capital*.

THE CAPITALIST AXIOMATIC

The Digital Future

For the return, then, of incommensurability in intellectual life.

Why Marx? And why Marx now? when most assuredly we have finally broken beyond the industrial stage of capitalism, finding ourselves either passive observers or active participants, but fully complicit in either case, in the unfolding destiny of the digital future, in the vectors of an emergent form of finance capital that through the medium of the digital nervous system and biogenetics is reprocessing the classical codes of the commodity-form into the flesh trade of surplus bodies and surplus imagination, and doing so in a traditional capitalist style that is global only in the sense of monopoly.

Perhaps the simplest response is that Marx can only really be read and critically appreciated now, in these specific historical circumstances when, more than a decade beyond the self-declared collapse of East Bloc 'official Communism,' Marx has lost his assigned role as theoretical ideologue for state socialism, reverting instantly into what his thought was always meant to be – nomadic political theory, certainly without an official residence, and, consequently, finally liberated to be critically retranslated in the historically specific conditions of the third millennium. Indeed, what could be better? Marx, the theoretician of the triumph of *Capital* writing in the midst of the turbulence of the industrial revolution, focusing his thought on the first great outbreaks of popular resistance, principally the Paris Commune and later the early political development of the international workers' movement, suddenly finds the pages of his texts – *Grundrisse, Capital, On German Ideology, The Eighteenth Brumaire of Louis Napoleon* – opened again to the early moments of the digital revolution.

Marx's thought rubs against digitality: wired culture's first popular resistances in the form of street disturbances against the closed future of the digital commodity, and its early political development of a potentially fatal contradiction in the triumph of virtual capitalism between knowledge-power and the increasingly closed circuit of (digital) circulation. Because Marx is no longer bound to state ideology, his thought is finally free to be thrown out of the theatrical productions of official power, to be forced into a new (critical) circuit of circulation. This migrant thinker of the dispossessed always was destined to be a political theorist of *Capital* in the nineteenth century and now a critical philosopher of *Streamed Capital* in the twenty-first. Out of time and certainly out of place, Marx, the first authentically critical thinker of the transnational political economy of (industrial) capital, now finds his thought thrown into the virtual circuit of circulation as a way of reporting in advance the class contradictions and alliances of global powers forming around the rebirth of the commodity-form in the triumphant language of the digital commodity-form. The hyper-realization of the market finds its (theoretical) equivalent in the hyper-realization of Marx. The fetishism of circulation is challenged from the grave by Marx's derealization of virtual capital. With Marx, the will to capital is simultaneously (historically) announced and (objectively) eclipsed.

A revived interest in Marx is always a register for other events, a sign of the pre-conscious presence of other historical symptoms. To return to Marx as a way of thinking the future is to admit the intellectual aridity surrounding the horizon of 'futurology' in the twenty-first century, and to speak of the impossibility, perhaps very real undesirability, of thinking one's historically proximate 'human condition' in the ruling language of those circumstances. In this sense, the exuberant language of technicity, whether expressed in the now superseded digital alliterations of 'e-commerce,' in the delirious visions of the wireless future, or in the virtually unstoppable drive towards a future of mega-(telecommunication)mergers, may be the shaping and informing language of the future that opens before us, but it is debatable whether the destiny of that future can be thought clearly in exclusively technical terms. Not that Marx was ever an anti-technologist. Quite the contrary. He was always the fully modern thinker, not so much a dark outrider of the Enlightenment as an indispensable *intellectual precondition* of the fulfilment of the modern project. Marx was a thinker of split consciousness: part pro-techne, part anti-capitalist. His thought was that critical margin of the Enlightenment project left to a more ancient, almost biblical, critique of the anima of capitalism while, at the same time, consenting to the modern drive to the fully realized technical future. In the long term, Marx paid the predictable price for his split consciousness: the privatization of (industrial) capital trumped his version of the socialization of technology. In the end, he was a thinker with nowhere to go, except to the slow speed of abandoned thought.

The Movement of Political Economy from a Strictly Capitalist Determination to Technicity

There is another story to be told about Marx. Not a repetition of the heroic theory of classical socialism with its historical struggle of workers against the bourgeoisie, use-value against exchange-value, but Marx viewed through the lens of quantum physics, specifically *uncertainty theory*. Here, it is precisely Marx's most critical failure – his failure to understand the *technology* in his *Capital* – that is made the starting-point for a theory of virtual capital that, in its broad critical lines, is a faithful translation of Marx's thought to the postmodern scene. Marx's thought as a field of uncertainty not totality, provisionality not absolutisms: critical only to the extent that Marx's *Capital* is intensified by paradox and irony as gateways to understanding the circularity of virtual capital. Consequently, Marx can be recuperated for the new century only by taking his (technological) aperçus seriously. Reversing his thought crystallizes its absences: critical aperçus that make possible Marx's present position as a brilliant guide to the political economy of virtual capitalism. Less the vanguard theorist of the nineteenth century, the Marx that reappears on the stage of contemporary history is theoretically rootless, drifting between a future of streamed capitalism and a past of command socialism, circulating between the speed of hyper-capitalism and the fatigue of state socialism.

And this makes sense. Marx was never the author of *Capital* in its most limited sense as a purely *economic* phenomenon. He never spoke simply of the fate of modern life in its strictly capitalist determination. Probably against his (political) intentions and certainly against his theoretical aspirations, he was always and only writing about the disappearance of capitalism into technology, the vanishing of a materialist theory of political economy into a *metaphysics* of the 'value-form' of *Capital* – the pure code of technicity.[5] Perhaps there is always this sense of a receding horizon in so many of Marx's writings. In *Capital*, *Grundrisse*, the *Economic and Philosophic Manuscripts*, no sooner does Marx lay claim to the material resolution of the enigma of capitalism – the 'fetishism' of the commodity-form – than suddenly capitalism slips from his fingers.[6] He is left declaiming that the *historical* expressions of capital – the material signs of labour and use-value, even the physical gestures of disciplinary capitalists – are just that, *illusive signs*, no sooner apprehended in the most *historical* of political economies than they disappear into something just as resolutely non-historical, something metaphysical, not the signs of capitalism, but the *'value-form'* of capital.

Could it be that capital as a 'value-form' – a semiological code governing the objectification of labour and the privatization of property – is really what is meant by economic manifestation of the 'will to will'? Is it possible that what Marx was really writing was the most recent chapter in a more ancient story: *the fantastic 'metamorphosis' of the will* – its critical and ubiquitous translation under the garb of modern capitalism from its strictly economic determination (the will to accumulation) to the will to virtuality? Wasn't Marx in the end only the latest, perhaps best, of the great classical metaphysicians? Marx, in fact, as a metaphysician of hyper-capital: a vision no less bold than Nietzsche's formulation of the will to power out of the psychological material of ressentiment or Heidegger's desolate imagery of a world abandoned by the gods to the reaping whirlwind of 'technicity.' Indeed, if the enigma of capital always fast receded before the probes of political economy that is because what Marx's thought really uncovers is that the constitutive grounds of political economy *never originate* in the logic of political economy proper, but in metaphysics. And not metaphysics in general, but in the metaphysics of the *will to virtuality* at exactly that point where the will to will – the will as simultaneously its own goal and its own constitutive ground of justification – abandons the theatre of representation, even the theatre of commodity-fetishism, contenting itself to be a self-referencing, self-validating, self-developing 'circuit of circulation.'[7] The pure value-form that is capital, therefore, as the value-form assumed by the 'will to will' in the era of streamed capitalism. The fatal realization of Marx's *Capital* is that capitalism is also a fetishistic sign of an *enframing* condition[8] – the appearance of the will to will in the form of a logic of pure circulation – that is preparing for its first open expression as world-history.

In this case, Marx's fate would be the traditional one of the sorcerer's apprentice calling into existence, however inadvertently, the conditions for the overcoming of capital, not however in the final interests of labour, but at the behest of the reign of the post-human. To surface the metaphysician in Marx, to ask questions concerning technicity as the constitutive ground of political economy, is to dig beneath the hygienic surface of his writing, bringing into presence the destiny of technology – *the logic of the value-form of capital* – that was always the magical elixir around which his thought hovered, and on account of which Marx's political theory has real surplus-value as the failure that deeply informs.

However, to ask if capital in following its seemingly preordained destiny has forgotten capitalism, to inquire into the metaphysical grounds of political economy, is ultimately unsatisfactory. Because let's be honest. What's really at stake is not only the psychological debt to a Nietzsche not yet written – the evil demon of ressentiment – that drives Marx on, making of his thought the Christian judgment on the capitalist times – no, what's really at stake is the *Heidegger in Marx*. Is Marx the fulfilment in advance of Heidegger's project of discovering the constitutive grounds of the 'will to will'? Is Heidegger's sense of 'fully technified life' simply a readout of the desolation following Marx's vision of virtual capitalism as an empty circuit of circulation? If Heidegger and Marx reach similar conclusions is this not because Heidegger is the truth-sayer of Marx, both of his successes and his failures, just as much as Marx is the impossibility (of specific historical commitment) that Heidegger could never overcome? To see the Heidegger in Marx is to radicalize Marx's thought by means of a metaphysics of streamed capitalism. To challenge Heidegger by Marx is to cut off Heidegger's retreat into the romance of Being, even dead being, in favour of leaving exposed like a raw digital nerve the question of what it means to live in a human condition typified by the boredom of forgetfulness, by the oblivion of being.

The Spectre of Politics

Returning to Marx also means recovering politics. The contemporary era is governed by the driving force of *command globalization*. Under the impact of a hyper-aggressive strategy of 'market penetration' organized around a bio-economic model featuring the cellular consolidation of the different sectors of traditional capitalism into the vast new media monopolies of virtual capitalism, the spectre of 'globalization' becomes *both the justifying condition and goal* of virtual capitalism. Here, just as Marx prophesied in *The Communist Manfesto*, the unfettered movement of the commodity-form breaks beyond the strictly economic sphere to involve the market penetration of every dimension of human experience, from electronically mediated subjectivity and processed (social) relations to the biogenetic engineering of human reproduction.[9] But unlike Marx's prophecies concerning the strict dialectical separation between the capitalist will to accumulation (under the sign of the privatization of property) and the will to labour emancipation (marked by the socialization of production), the globalization of the high-intensity (digital) market setting penetrates the old dialectic of privatization and socialization with such relentless force that globalization quickly comes to represent both the socialization of (virtual) production and the privatization of (collective) labour. Not so much Baudrillard's 'rational terrorism of the code,' but the cold seduction of the will to virtuality. The global political economy of virtual capitalism opens to reveal the hyper-reality of a networked planet reconfigured by a triumphant logic of virtualization, with quick flips between acquisitiveness and boredom as its dominant signs.

Streamed capitalism, then, as a culture typified by the ecstasy of bleakness in the midst of panic stampedes to cyber-business. Like the gold rush that marked the transitional moments between the end of the nineteenth century and the beginning of the twentieth, the beginning of the third millennium is typified by a digital rush out of material production into the virtual logic of streamed capitalism. The labour apparatus sheds workers; banks transform real material users into 'clients' serviced remotely by financial consultants; education remodels itself on the corporate model of 'e-students'; and business transactions take place increasingly in the realm of the electronic. Here, streamed capital announces its presence, at first in the now failed language of e-com ventures and the exuberantly priced shares of cyber-businesses at the forward edge of the wireless future, and then, once purged through a worldwide recession, in terms of the global consolidation of multinational corporations into *branded electronic networks*, not domiciled in a fixed geographical location, but representative only of a *strategic node* in the circulation of the digital circuit. Streamed capitalism, therefore, as a dynamic vector populated by a global multitude of increasingly isolated wired individuals, driven forward by alternating currents of wealth and necessity, and organized into the new electronic planets of streamed multinational corporations. For those self-selected to be servomechanisms of the electronic future, there is a reality of hyper-fatigue jobs in return for the illusion, and sometimes reality, of unfathomable wealth. For those not involved in a strategic digital node, there is only a growing sense of frustration and humiliation at having been left behind, coupled with the pressure of labour discipline in a recessionary economy. Chronic dissatisfaction and fading dreams of escape then, as the twin cultural poles of life in the digital future.

Hyper-Capitalism

The impossible odds of contemporary political history, this strange field of business hyper-reality where vast multinationals metastasize instantly in a doubled logic of connectivity and control and individuals are compressed into digital terminals, makes Marx interesting once more. Marx's critique of capitalism was always premised on taking capitalism out of the cycle of (economic) production and putting it into *political* circulation. His was a strategy of political excess that undermined the laws of commodity-production by means of a competing theatre of political representation. His voice was marked by the sounds of a more ancient physics of the human spirit: the crushing solitude of private acquisition versus collective emancipation, extractive labour-power versus the intrinsic value of labour in (human) use, accumulation versus justice – in short, *economics versus politics*. To speak the name of Marx is to breathe again the air of political rebellion, and this is probably why, more than a century after his death, the name of Marx is still a sign, a talisman, and a 'spectre' of political resistance.

Always advocating a politics of excess, Marx diagnosed the logic of capitalism with such relentless force that it was compelled to reveal itself as having, in the end, only an *incidental* relationship to production, but a *necessary* relationship to capital understood as a 'circuit of circulation.' Moreover, Marx theorized the politics of opposition to capitalism in its first appearance as a 'mode of production' with such precision that it might be said that the most radical politics of the modern century were staged by his political playwriting. Capitalism versus communism, liberalism versus socialism, workers versus bourgeoisie, radical theology against the compromised logic of the Catholic hierarchy: all a theatre of political representation performed in Marx's shadow. His writing represented a continental fracture in the order of capitalist existence.

From Capitalism as a 'Theatre of Representation' to Virtuality

If streamed capitalism replaces the theatre of capitalist representation with virtuality, then the return to Marx also puts into question the order of politics pertinent to the new regime of virtual capitalism. When digital capitalism shifts from labour-power to knowledge-power, from the commodity-form to the circulation of *the digital circuit itself* – then what of political opposition?

Streamed Marx, therefore, for the era of virtual capitalism.

The Virtual Class[10]

The Marx that belongs in the streets of the twenty-first century is also one that theorizes politics in terms of its actual historical predicament, namely the spectre of globalization. If the expression of political interests necessarily follows the interest of economic expression, then the politics of virtuality brings into existence a *new class*: a class with no previous collective identity. A *virtual class* which, forcibly breaking with the mode of (industrial) production, quickly aligns itself as the class representative of the digital commodity-form. The virtual class is global, networked, liquid, connected, controlling, and fungible in its technical labour skills, a specialist class of the digital nervous system. It is simultaneously the historical successor to the declining class of the old bourgeoisie – the privileged owners of the means of capitalist production – and its epitaph. Not simply a direct representative of capitalist interests but, instead, a virtual sign of the *political alliance* necessary to commence streamed capitalism, the virtual class is an objective realization of the hyper-reality of the new economy. Here, the many specialists with a vested interest in maximizing the digital nervous system congeal into an increasingly unified new global class: venture capitalists of the wireless world, owners of new economy stocks, IPO entrepreneurs, software CEOs, multimedia designers (with a falling rate of options), digital consultants, marketing and encryption experts, Web designers, new media directors, intellectual property rights law firms. As the decisive *knowledge factor* in the new knowledge theory of value, as the *collective designer* of the digital mode of production, as the *essential intellectual circuitry* in the circulation of digital circuits, as the *key aesthetic node* in the cultural economy of the new media, the virtual class carries out the necessary historical task of materializing the digital nerve as the key code of virtual capitalism. Unconscious of its broader historical role, always engaged in ruthless competition, here bringing together the fields of digitality and biogenetics, there projecting the codes of technology into every dimension of human experience, subordinating the state to the ideology of technicity, brilliantly interfacing money and genius, the virtual class finally realizes the ascendant logic of streamed capitalism as the triumph of virtuality. The virtual class is *how* the circulation of the digital circuit is deep-time coded into every dimension of human experience. The virtual class *harvests* the world for the digital future. It is a Heideggerian class working in the practico-inert of Marxian social history.

FOUR THESES

So then, four theses on the unfolding destiny of the capitalist axiomatic in the twenty-first century: an *economic thesis* – the *circulating commodity*[11] as the essence of the immediate process of virtual production; an *epistemological thesis* – the *knowledge theory of value* (intellectual property rights) as the crucial axiom of the circuit of digital production; an *aesthetic thesis – virtual value* in the transcendent form of 'value valorizing itself' as the key value-form of virtual capitalism; and, finally, a *historical thesis* – the *disciplining of capitalism* by the planetary drive to technicity. Less a reprise of Marx's *Capital* than an intensification of Marx's political economy in light of the contemporary hyper-realization of the marketplace and the fetishism of circulation, these four theses concerning the capitalist axiomatic are intended to write the final postscript of *Capital*, to take Marx seriously when he spoke so evocatively of capitalism as a 'circuit of circulation,' of the necessary alternation of all positions within the code of the capitalist axiomatic, to say finally with Marx that while a material theory of history is the surface sign of *Capital*, this book is above all a metaphysical text, a text which, refusing to end in the terminals of labour or bourgeoisie or in that of political economy understood as providing its own ground, speaks eloquently and well of a phantom presence haunting the presence of *Capital*, the phantom of speed, which once freed from production transforms itself immediately into the spectre of circulation, into a networked world moving at hyper-acceleration. Could it be that the real spectre haunting the history of *Capital* is not that of use-value, of the almost theological affiliation of labour with the stability of referential value, but the transformation of the dialectic of use-value and exchange-value into a *third term* – the spectre of an empty cycle of circulation moving at the digital speed of degree-zero and in circuits of zero-time? In this case, the spectre of *Capital* would be *Virtual* Capital: the creation out of the historical ruins of *Capital* of a dynamic vector of virtual capitalism: a single, hyper-dimensional, global economic process driven forward by the commodification of circulation with such intensity that the new mode of production – *digital production* – ushers in a qualitatively new historical epoch typified by knowledge-power not labour-power, virtual-value not exchange-value, and the transformation of capitalism in the direction of a culture of technicity not surplus-value.

1st Axiomatic: The Speed of Circulation

The Immediate Process of (Virtual) Production

Written as an appendix to *Capital*, 'The Results of the Immediate Process of Production' is Marx's manifesto for the liberation of the commodity-form from its understanding as an 'autonomous object' invested with use-value to the critical moment in the metamorphosis of Capitalism itself.[12] Here, in a series of specific theorizations, Marx diagnoses the ruling axiomatic of industrial Capitalism: 'Commodities as a product of Capital'; 'Capitalist Production as the Production of Surplus-Value'; 'Capitalist Production Is the Production and Reproduction of Specifically Capitalist Relations of Production.' If 'overcoming Marx' means not abandoning Marx as a brilliant theorist of a now derealized form of Capitalism, but intensifying Marx's insights into the 'immediate process of capitalist production,' then this series of theorizations is a critical gateway, simultaneously into Marx and beyond him. Finally we are in a position not only to anticipate in advance, but to diagnose, the real, *virtual* results of the immediate process of production. Overcoming Marx, therefore, by intensifying *Capital* so as to yield *the results of the immediate process of* virtual *production*.

The Circulating Commodity

And why not? If Marx can have such a continuing animating quality, if his writings continue to disturb the ruling capitalist class so deeply, perhaps it's because Marx was never purely a theorist of the industrial stage of Capitalism, but was the first theorist of digital Capitalism. For this is the crucial moment of 'The Immediate Process of Capitalist Production.' It is not really about Capitalism as production, but about Capitalism as circulation. In Marx's estimation, the charisma of Capitalism rests on the fact that the Capitalism is always 'in circulation,' only episodically coming out of circulation to enter into production. That's the role of the commodity-form. It performs the aesthetic function of a *hinge experience*:[13] injecting itself into the world of raw materiality – human labour, agriculture, manufacturing – only to immediately recode material experience with itself as the new 'universal, elementary form of production.' While the commodity-form pre-dates Capitalism (Marx discusses the commodity as also existent in pre-bourgeois modes of production), the commodity-form has this particular quality. It is a cybernetic code, simultaneously transforming material experience into a 'universal, elementary form of value' and representing the possibility of 'transfiguring' use-value into exchange-value, constant Capital and variable Capital into surplus-value. The commodity-form is the primary cybernetic code in Capitalism understood as a process of 'valorization.' In its first appearance as the universal value-form of the process of industrial Capitalism, the commodity-form is a sign-form.

While the writings of Marx have themselves a 'distinct social character' invested with all the millenarian dreams of socialist resistance to the Capitalist axiomatic, what drives Marx's thought forward is his profound insight into Capitalism as 'pure circulation,' as only provisionally, but not necessarily, about production. Here, the process of Capitalist valorization signalled by the appearance of the commodity-form as the 'universal, elementary form of production' finds its first provisional language in the unification of labour-power and the valorization process, in the metamorphosis of the process of Capitalist production into the form of Capitalist accumulation. But the first historical appearance of the commodity-form as the 'Capitalist depositary'[14] of the immediate process of production is always and only a first tentative step towards the disappearance of the 'immediate process of production' into the production of the process of immediate virtualization. For Marx, capitalism as a relentless 'circuit of circulation' ultimately overthrows the commodity-form understood as the universal, elementary *value-form* of the immediate process of (industrial) production in favour of the *circulating commodity*,[15] the commodity that never leaves the circuit of circulation, never enters directly into the process of production, never ceases to be the universal, elementary form of the valorization process. In its latest appearance as the value-form of the digital phase of Capitalist production, the *commodity-form is a virtual form*. Abandoning production in favour of speed of circulation, disappearing labour at the behest of Capitalism that increasingly assumes the form of an artificial intelligence construct, overcoming the dialectics of use-value and exchange-value in favour of the virtualization of the value-form, the digital commodity-form assumes the universal, elementary form of the stock market. Here, the 'circuit of circulation' that was always the repressed dream of Capitalist production can finally achieve determinate (virtual) form. The stock market, then, as the fully realized expression of the digital commodity-form: pure commodity, pure value, pure speed, pure code, pure virtuality, pure price.

The Mirror of Velocity

As the universally necessary form of the product, as the specific characteristic of the process of production, the commodity palpably comes into its own in the course of Capitalist production. The product becomes increasingly one-sided and massive in character.[16]

Whereas in industrial Capitalism, the commodity undergoes a transformation in which the 'product assumes its commodity-character and hence its exchange-value,'[17] in digital Capitalism it is the commodity itself which assumes its virtual character and hence its virtual value. In industrial Capitalism, the commodity unifies labour-power and the valorization process. In digital Capitalism, the process of valorization unifies knowledge-power and the speed of circulation of the digital network. In the manufacturing stage of Capitalism, the commodity assumes an immediate 'social character' and is 'one-sided and massive in character.' In digital Capitalism, the digital commodity has a distinct virtual character, a vector not a magnitude, n-dimensional not one-sided. Here, the valorization process is about speed not labour, virtuality not sociality. It is about the character of circulation, not the circulation of the social character of the commodity-form.

As to the question, 'What is objectified in the digital commodity-form?' the response: not labour (except as a residual category), *but knowledge*. A 'Capitalist depositary,' the digital commodity-form objectifies knowledge to such a degree of intensity that we may speak now not of surplus-labour, but of surplus-knowledge – cybernetic knowledge in the form of 'intellectual property rights' expressive of the digital nervous system which, breaking free from its ground in individual consciousness, floats across the spectrum of digital nodes as a pure circulating medium of digital exchange..

Marx was correct. 'The commodity is a transformation of Capital that has valorized itself.'[18] But whereas in the industrial stage of Capitalism, Marx located the code of the commodity in a valorization process wherein 'the commodity must acquire a twofold mode of existence if it is to be rendered fit for the circulation process'[19] (use-value and exchange-value), in digital Capitalism the virtualization of the commodity-form injects a third term into the equation. No longer the dialectic of use-value and exchange-value, but now a *recombinant commodity-form* wherein the speed of circulation depends for its very existence on the transformation of use-value into an economics of speed and exchange-value into an economy of virtualities. Whereas what is valorized in industrial Capitalism is the commodity-form itself as the 'universal, elementary value-form,'[20] in digital Capitalism it is Capital itself that is valorized in the form of singularities and networks and portals and gateways. When the value of the commodity-form finds its equivalence in the speed of circulation, then the circulation of the commodity-form of value is the pure sign of (digital) Capitalism that has succeeded in valorizing itself. Virtual Capitalism, then, as a mirror of velocity.

From Labour as a 'Factor of Production' to the Production of 'Factored Labour'

> Hence the rule of the Capitalist over the worker is the rule of things over man, of dead labour over the living, of the product over the producer.[21]

Marx claims that the result of the Capitalist mode of production 'is steadily to increase the productivity of labour.'[22] But what if under the pressure of the speed of circulation of the Capitalist axiomatic associated with the digital commodity-form, the 'productivity of labour' were to be increased with such intensification that labour itself as a 'factor of production' is disappeared in favour of the production of factored labour? Factored labour? That is digital labour not as a purely economic phenomenon, but as a relentless exercise in *political discipline*. Digital labour as a circulating medium of networked exchange – coded, portalled, networked, and mediated – that is simultaneously a precondition of the speed of circulation of the digital commodity-form and the stamped product of the digital nervous system. Factored labour, then, as a regime of *disciplinary managerial practices*: a coded set of work procedures the goal of which is to create a seamless interface of human flesh and machines. In this case, factored labour would materialize in the work situation that Marx prophesied in *Capital*: a future not of 'living labour' but 'dead labour.' Dead labour, that is, as an artificial construct of managerial procedures tattooed on the labouring body.

This is the real meaning of virtualization. Not simply the virtualization of Capital by the movement beyond production to the valorization of the circuit of Capital itself, but throwing labouring flesh into the speed vectors of virtual Capital. Here, beyond Marx's understanding of work as essentially involving the private appropriation of the surplus-value of living labour, what is really appropriated in digital Capitalism is the time of human surplus itself: the relentless translation of every gesture of the labouring body into an active reinforcement of the Capitalist axiomatic. Beyond psychological repression, virtual labour concerns the shutting down of living, embodied labour, and the triumph of human flesh as a disciplined labour machine. This is also Heidegger's language of harvesting: our reduction to the inertia of the 'standing-reserve.'

2nd Axiomatic: The Knowledge Theory of Value

From Splitting the Atom of Labour to the Virtualization of Knowledge

Always the question of labour: the labour theory of value, the political economy of labour, the fateful division of labour and labour-power, surplus value extracted from the labour of working men and women, labour as the material pole of use-value and the regulative sign of the Capitalist exploitation of exchange-value, labour simultaneously as the sign of universal human emancipation and the mark of despotic Capitalism tattooed on the skin of the proletariat.

To overcome Marx, that is, to appropriate what is most creatively intense about Marx for an understanding of the Capitalist axiomatic in the digital era, it will always be necessary to return to the question of labour. To return, however, not by way of a frontal embrace that seeks to reinstall Marx's understanding of labour power as the key to digital Capitalism, but to return by investigating, as would Marx himself, what is the actually existent condition of labour and labour power in the cybernetic era. That this is a fatal question has already been suggested by Marx himself, since he stopped long ago being a political economist and became a proletarian semiologist. Probably the first. When Marx first split the atom of labour, when he distinguished between the particular subjectivity of labour and the universal value of labour power, when he first intimated that nested deeply within the material tissue of labouring flesh was the doubled sign of labour waiting only to be recognized before it would attempt to make its claims of universal sovereignty in world history, well when Marx first split the atom of semiotic labour, when he performed his own Nietzschean transvaluation of labour-value on behalf of a really existent theory of labour-power, at that point Marx himself subordinated the romance of the labouring body to the cybernetics of labour power.

Because that is what Marx's theory of labour-power is: a critical theory of post-labour. That point where the specificity of labouring bodies is subordinated to the value-form of generalized bodies of labouring conditions. Before the labour theory of value, there existed, however untheorized, real material labour in definite historical conditions of use and abuse. After the promulgation of the labour theory of value, there was labour-power as the ruling axiomatic of a cybernetic duel of use-value and exchange-value. Consequently, with Marx the quintessential theorist of labour, this ironic outcome: the philosophical defeat of the incommensurability of labouring activity, and the triumph of the universal form of labour value.

Digital Sweat Shops as Marx's 'Dead Labour'

Now, many working theorists have been this way before. Jean-François Lyotard recognized the death knell in the war machine of labour power, immediately calling for a recovery of a 'libidinal economy,'[23] a primitive pre-Marxism that would take labouring flesh back to an economy of pleasure and pain, an economy of sex and muscle, with the psychoanalytics of the suppressed id as the wild card in between. In *The Mirror of Production*, Jean Baudrillard was less utopian, more severe.[24] In a twentieth-century reprise of that earlier meeting of French revolutionary thought of 1848 with the philosophy of the German Enlightenment, Jacobins with Hegel, the Paris Commune with German conservative reactionism, Baudrillard diagnosed Marx's sleight of hand at the splitting of labour and labour-power for what it was: the foundations of a 'structural law of value'[25] in which Capitalism was to be cyberneticized in the form of the ruling axiomatic of signifier (use-value) and signified (exchange-value). More than Lyotard, who placed his theoretical bets on drawing out the differend in Marx's theory of labour, Baudrillard wrote *The Mirror of Production* as Marx's avenging angel, always insisting that the labour theory of value, far from inaugurating a new possibility of proletarian, let alone universal, emancipation, only functioned to install the 'rational terrorism of the code'[26] as the ruling order of the Capitalist axiomatic. For his fateful insights into the happy complicity of Marxism and Capitalism in sharing exactly the same episteme of the structural law of value, Baudrillard has suffered the usual fate: denounced everywhere as a cynic, this the most political of all thinkers, the most rigorously and naturally Marxist of all the post-Marxists, banished to an early twilight of thought on the (Parisian) margins.

If the twin gestures of Lyotard's *Libidinal Economy* and Baudrillard's *The Mirror of Production* cannot be assimilated into conventional Marxism, perhaps that is because Marx himself is the first post-Marxist, the first theorist, that is, who, unsettled by the ambiguities of identity and difference disguised here as the language of labour and labour-power, there as the supposedly warring poles of use-value and exchange-value, wrote out his labour theory of value as a secret prolegomenon to a stage of Capitalism not yet existent. A code waiting for a material form, a theory of value awaiting its definite historical conditions of labour. Marx's *Capital*, then, not so much an analysis of the definite conditions for the socialization of labour, although that too, but a secret history, written much in the tradition of the cabal, of a future Capitalism that would represent, as Marx theorizes in the third volume of *Capital*, the really existent history of 'dead labour.' It is, of course, around the corpse of Marx's 'dead labour' that *Libidinal Economy* and *The Mirror of Production* circle, sometimes settling in the desert sands for a feeding frenzy of delirious theory, sometimes startled by the approach of passing armies of the night, taking flight, almost like a permanent evocation of ecstasy that was the Paris Commune and the slow decline of the German bourgeois restoration. Whether by way of Marx, who studiously wrote out the codes of the (industrial) Capitalist axiomatic only to write 'dead labour' as its epitaph and future, or by virtue of the chilling lightness of being of the French mind in the form of Lyotard and Baudrillard, who always and only begin with the last rites of dead labour, the spectre of post-labour haunts the Capitalist axiomatic.

And this is as it should be. Because whether expressed in the language of the 'structural law of value' or in terms of the 'labour theory of value,' Capitalism has never been about really existent labour, but always about the subordination of labour to the theory of value. Modern Capitalism, then, as an industrial semiotics. A critical point because it indicates that if contemporary Capitalism – digital Capitalism with its globalized markets and circular flows of Capital – can so easily escape the question of labour, can so swiftly disappear labour in the (domestic) form of 'reductions in the work-place' or in the (international) form of harvesting the world for perpetually new sources of cheap labour, that does not indicate that Capitalism has escaped the question of value. So then, two alternative theses. First, an *epistemological* thesis concerning the historical transition of the value-form of Capitalism: modern Capitalism might have been coded by the labour theory of value, but twenty-first-century Capitalism will be organized under the sign of the knowledge-theory of value. Digital Capitalism as networked knowledge, not a labour exchange. Certainly the safest (theoretical) route. It requires for its affirmation only an acceptance of the received wisdom that the epochal movement from industrial to cybernetic Capitalism, from modern Capitalism premised on the labour theory of value to postmodern Capitalism based on the knowledge theory of value, takes the form of a general historical movement to a knowledge-based society, with all its implications for the triumph of a class of networked specialists over traditional labour based in fading manufacturing and service sectors. Or, an opposing *ontological* thesis. A strange thesis concerning the ambiguity of labour and knowledge, that labouring activity is always knowledge-based activity, that the knowledge-based society has its deepest genealogical roots in the first worker who mastered the cybernetics of the machine age, and that if the fantastic knowledge of the worker was somehow lost sight of in the rush to the labour-theory of value that is not to say that it was not always labour-value, but knowledge-value, that was put into play by the Capitalist axiomatic. Out of this latter thesis, there emerges an enigmatic Marx. Not Marx as the nineteenth-century theorist of the approaching struggle of proletariat and bourgeoisie over the alternatively exploitative and emancipatory rites of the labour theory of value, but the spectre of Marx as the twenty-first-century theorist who, in diagnosing the objective historical conditions for labour-value, was in reality, a hundred and fifty years before its time, analysing the future-world of an advanced Capitalist society premised on the knowledge-theory of value. With the triumph of digital Capitalism, it may well be that the proletarianization of knowledge-work is only about to begin. If this is the case then Marx's theory of class struggle may yet be awaiting its first moment of historical realization in a fateful struggle over the falling rate of (digital) profit and the exploitation of knowledge-value. From the exploitation of industrial labour to the struggle over knowledge-value, understanding the digital future is in its Marxian past.

From Dead Labour to Virtual Knowledge

Digital Capitalism is focused on the struggle over the emancipation of knowledge-value from embodied knowledge: its genealogy, its circulation as a generalized medium of cybernetic exchange, and its final appearance, in the residual form not of dead labour, but of dead knowledge.

In exactly the same way that labour functioned in industrial Capitalism both as an embodiment of use-value and as the dominant value-form of exchange-value, knowledge functions in digital Capitalism as a regulator of use-labour and exchange-labour. The dominant medium of exchange of digital Capitalism, knowledge breaks forever with its genealogical roots in individual consciousness, becoming instead the key value-form of cybernetic culture. Not living knowledge with its origins in individual subjectivity, embodied memory, or granulated knowledge of specific labour practices, but virtual knowledge as the dynamic medium of cybernetic culture.

Virtual Knowledge? That's knowledge when it is first invested with the alien quality of value, when knowledge first ceases to a matter of individual consciousness, becoming instead a value-form, a circulating medium within whose deflationary and inflationary flows the world is cyberneticized, coded here with the binary logic of 0/1, there spoken of as a 'digital nervous system.' In the same way that labour was a measure of market value in modern Capitalism, knowledge is now a value of the measure of cyberneticity of digital Capitalism. It is around the question of knowledge-value that the fundamental class conflicts of digital Capitalism first organize, sometimes in the daily repressed form of the domination of human imagination by the performance-requirements of the virtual workplace, at other times by open rebellions of a rising tide of global counter-knowledge – environmentalism, collective labour rights, human rights – against the 'free circulation' of knowledge-value, which is to say of cybernetic Capital, so necessary for the continued functioned of digital Capitalism.

Far from abandoning the generic pattern of use-value and exchange-value so endemic to the industrial mode of production, digital Capitalism perpetuates, even intensifies, this pattern, but at a higher level of abstraction and generality. Severing the bond of use-knowledge and individual subjectivity, digital Capitalism codes the question of use-knowledge in its deepest interiority. What once may have accurately been described as psychological repression now takes the form of a gradual constriction of knowledge to a doubled code of use. On the one hand, knowledge that is instrumentally useful for the seamless insertion of human flesh into the high-performance speed of digital culture, and on the other, codes of use that are only apparently transgressive. Virtual knowledge and abuse knowledge, both contained as simultaneously explosive and domesticated tendencies in the same body. 'Be serious' and 'be fake' as the doubled sign of the knowledge-worker fully prepared for fast integration into the axiomatic of digital Capitalism. With instrumental (digital) knowledge, the body is plugged into the disciplinary network of careerism. With (cultural) abuse value, the body touches at a distance that which has been forgotten in the interfacing of human and machine flesh: libidinal pleasures, human memory, symbolic exchange. All these are simultaneously retrieved in a (cultural) way that immediately shuts down the anxiety of that which has been forgotten. It is 'be fake' culture: the appropriation of the original ecstasy spirit of rave culture by the corporate cultural axiomatic, the assimilation of sometimes ghetto, sometimes prison, rap culture for the ersatz entertainment of white suburban boys. Economically, use-knowledge in the digital axiomatic binds human flesh to the digital nervous system. Culturally, digital use-knowledge is about the cynicism of forgetfulness.

Connectivity and forgetfulness: that is the paradoxical condition of virtual knowledge in the ruling capitalist axiomatic.

3rd Axiomatic: The Virtual Class as the Objectification of 'Value Valorizing Itself'

Differing sharply from the bourgeoisie of traditional Capitalism with its exclusive ownership of the means of production, the virtual class virtualizes the production of the means of intellectual ownership. No longer the possessors of private property, the virtual class is now the owner of the means of virtual property – intellectual property. This change is historically decisive. While private property is grounded in the always antagonistic relationship of labour and property, intellectual property is about the exclusive exploitation of the knowledge theory of value. Here, knowledge is the exclusive medium of intellectual property, and intellectual property – its creation, coding, patenting, and distribution – is the motor-force of the digital commodity-form.

Indeed, the digital commodity-form takes the immediate form of the production, exchange, distribution, and consumption of intellectual property rights – the contemporary economic code scripting exclusive knowledge about the patenting of the digital future. Once effectively separated from embodied consciousness, virtual knowledge in the privatized form of intellectual property rights quickly emerges as the disembodied nervous system of digital reality. In relationship to intellectual property, the virtual class is a form of virtual midwifery – theorizing the future of the digital nervous system, creating its code-structure, designing its architecture, always driven on by the discipline of Capitalist competition to create turbulence in the digital system, here restricted to imagining the digital future in terms of a solely Capitalist determination, there reimagining the human future through the prism of an almost spiritual sense of the digital geist. In some nations, intellectual property-workers are passive servomechanisms of digital colonialism. In others, the virtual class is at the creative (imperial) centre of digital coding. However, in all cases, the virtual class is the leading intelligence of the digital nerve, creating, designing, exchanging, distributing, consuming, and finally, ensuring the legal and extra-legal hegemony of the sphere of intellectual property rights as the source code of virtual capitalism itself.

Virtual Value as the *Virtualization* of the Value-Form of Capitalism

In the global economy, corporations today have value only in relationship to the virtualization of the value-form of Capital. In traditional Capitalism, profitability is the real measure of value, and long-term solvency the code-principle for assessing the risk ratio of financial Capital. In digital Capitalism, solvency is projected into the future as a virtual expectation of successful Net corporations. Here, virtual value is the measure of the long-term solvency of (electronic) profitability, and profitability the alibi projected into the future under the sign of the illusion of solvency of digital Capital. Breaking with the modernist axiomatic of Capital where the restless migration of Capital was regulated by the relationship of (profit) risk to economic solvency, the digital axiomatic undermines (present) profitability and with it traditional financial theorizations of the relationship of price/earnings ratio in favour of the probability of favourable market positioning in the digital future. Indeed, to the extent that digital corporations show an actual profit as measured in terms of dividends, it probably only indicates that they are still rooted too deeply in the code-principles of the (modernist) Capitalist axiomatic, still too wedded to the solvency of the sign-value of Capital, not yet emancipated from Capital as a sign-value to the (postmodern) axiomatic of Capital as a virtual value.

In the new economy, the stock market with its delirious dreams, vectored flows, and fast rise and fall of share pricing is a decisive measure of virtual value. Here, solvency gives way to the probability of future (digital) market penetration, actual profitability is subordinated to minute measurements of (virtual) expectations of quarterly earnings, and the digital corporation as a symbol of the bureaucratic management of Capital is quickly replaced by the image of the Net corporation as a self-regulating, self-reflexive platform of software intelligence providing a privileged portal into the digital universe.

Between Technology and Capitalism

The fate of the virtual class is inextricably tied to the knowledge theory of value. As a class representing the leading (theoretical) edge of the new economy, the virtual class simultaneously brings virtual knowledge into real (digital) material existence, patenting its ownership rights, codifying its procedures of practical (net) use, and recreating the world in terms of a three-tier virtual architecture. At the same time that the virtual class is the self-realized form of the knowledge theory of value, the very existence of the virtual class is itself dependent on the actualization of virtual knowledge. Simultaneously a decisive point of the historical realization of virtual knowledge as well as its world cypher, the virtual class always occupies the dual position of (active) creator and (passive) portal, explosive dynamic of the digital future and servomechanism for consolidating the digital past.

But at the same moment that the virtual class materializes the knowledge theory of value in terms of a radical remaking of every dimension of existence in the direction of pure technicity, at the exact point that the virtual class represents the real material realization of the will to virtuality, at that specific historical juncture the virtual class inserts into the brainware of the digital nervous system a fundamental objective contradiction that has the effect of forcing the will to virtuality to undergo a fatal oscillation of its wave-form. Never a class capable of being subordinated either to pure technological (use) value or to relentlessly accumulative Capitalist (exchange) value, the peculiarity of the virtual class is that it stands midway between technology and Capitalism. It is the inherently unstable third term mediating Capitalism and technology, sometimes their point of reconciliation as expressed by the vanguard (networked) multinationals of the new economy, at other times their moment of internal disturbance as represented both by the speed with which new (electronic) technologies undermine settled Capitalist formations and the discipline with which the drive to economic profitability objectively limits technological creativity. Always a dynamic mediation, always a restless 'going across' between the imperatives of technology and Capitalism, the virtual class simultaneously represents the cybernetic intelligence necessary to realize the will to virtuality and its potentiality for a fatal undermining.

Between technology and Capitalism? That's the virtual class in its purely aesthetic mode as a 'hinge' experience, here attaching itself to established Capitalist formations as a way of gaining access to the technological apparatus, there practising a form of (cyber) monastic self-discipline as it shrinks its collective technological imagination down to the instrumental size of market-exchange. However, the virtual class can never remain for long under the spell of a strictly Capitalist determination because it is, first and foremost, as a virtual class the historical embodiment of the will to virtuality. While necessity grafts it to Capitalism, its own (technological) class imperative forces it to continuously upgrade itself to the advancing edge of the cybernetic storm. Always post-Capitalist, pure technicity destroys national structures of Capitalism in favour of the virtualizing imperatives of globalization. Always post-accumulative, the will to virtuality is imminently disaccumulative, replacing the sign-form of Capital with the value-form of technology and, thereby, reducing the laws of market exchange to the self-regulating discipline of cybernicity. Always post-market, the new economy is an anti-market, replacing consumption with pure (financial) exchange, market instrumentality with crowd contagion, the signs of a ruling economy with the ruling economy of cultural fads. Standing between Capitalism and technology, the virtual class has split consciousness, and in the growing consciousness of this split it is the difference that simultaneously realizes the will to virtuality and derealizes the virtualization of the will to technology.

4th Axiomatic: The Disciplining of Capital by Technicity

'Alien Value'

Or is something else at work in the contemporary era? Not simply the Capitalists' ability to supervise and discipline labor,[27] but just the opposite, technology's ability to supervise and discipline Capital. Just as the Capitalist 'will always seek to extend the duration and intensity of work,' the will to virtuality necessitates the extension of the duration and intensity of Capital. The process of Capitalist production is not allowed to pause: consequently, the twenty-four-hour stock market pacing the celestial movements of the earth as it rotates around the sun, after-hours trading as a way of disturbing the rhythms of daytime trading, streamed capitalism as the final triumph of extensiveness over duration, of dead (technological) space over the time of living Capital.

Whereas in the industrial Capitalism of the twentieth century, the decisive political struggle was between labour and Capital, in the virtual Capitalism of the twenty-first century, the key historical contradiction lies between Capital and virtuality, between, that is, the will to Capitalist accumulation and the accumulative instincts of the will to virtuality.

In a fantastic intensification and abstraction of the historical drama between living labour and dead Capital charted in all its theoretical precision and economic granularity by Marx's *Capital*, virtual Capitalism means, in its essentials, that in the twenty-first century Capital will, and does, occupy the same position as labour in the modern era. Just as the worker in relation to the Capitalist appropriation of surplus-value creates a 'value alien'[28] to himself, namely the valorization of the Capitalist process of production, so too Capital in relationship to virtuality also creates a value alien to itself. Incorporated into the digital process of production, at first as its necessary condition of historical development and later as its spectacular product, Capital now creates a surplus-value of virtuality, a speed and magnitude of digital technology that is not reducible for its explanation to the laws of Capital accumulation, and on behalf of which Capital is forced to serve as its historical incubator. While in industrial Capitalism, Capital in its fateful contestation with labour was the leading historical force, in virtual Capitalism, Capitalism declines into a residual historical category, a transitional state between the historically particular absorption of labour into the machinery of Capitalist accumulation and the global appropriation of Capitalism into a dynamic medium for the valorization of the technological process of (digital) production.

The Genealogy of Virtual Capitalism

Consequently, the great value of reading Marx's *Capital* retrospectively, not from the vantage-point of the modern century on the cusp of the supposed post-Capitalist future but from the perspective of the much-vaunted digital future looking backwards, is that Marx provides both a brilliant genealogy of the birth of Capitalism and an equally insightful 'futurology' of its eclipse. Not, however, the eclipse of Capitalism by its 'objective' historical opponent – living labour – but the overcoming of Capitalism by its historical transformation into its pure (metaphysical) form, what Marx glimpsed from afar as the 'circuit of circulation' and what we now experience directly as the will to virtuality. Shedding the necessity of actual production, transcending the model of consumption, transforming distribution into networked vectors, and accelerating the language of exchange into mathematics of a 'three-tiered digital architecture,' the will to virtuality exists now as the pure form of the circuit of Capitalism that Marx first prophesied, and which in its first tentative steps announces itself, at first hesitatingly and then gathering strength from its global momentum, as the triumph of the will to virtuality.

Relentlessly focused on the political struggle of Capital and labour, socialization and privatization, collective consciousness and individual accumulation, Marx's thought is primarily about labour, only secondarily about Capital, a matter of dialectics, not code. Thus the great irony. It is not to the dialectics of living labour and dead labour that we are compelled, finally and critically, to look for the sources of the overcoming of Capitalism, but to the metaphysics of Capitalism itself. Capitalism will not, and does not, survive its transformation from content to form, from its material-form to its value-form, from a circuit of production to a pure circuit of circulation. And why? Because Capitalism as a 'circuit of circulation,' in its contemporary representation as a pure form of valorization, is only apparently the value-form of Capital accumulation, but concretely the value-form of virtuality. At the heart of Marx's *Capital* is less a haunting analysis of the dialectics of labour and Capital than a prophetic diagnosis of the hidden code of virtuality. Because that's what Marx's language of the 'value-form' of Capitalism really concerns, not Capital accumulation in its strict economic determination but the accumulative energies of the 'circuit of circulation' in its dynamic technological determination. Set in motion by the historical alibi of market-driven Capitalist accumulation, by the market-subordinated language of production, consumption, distribution, and exchange, the 'circuit of circulation' cannot, and does not, resist valorizing itself as its own value-form. Not the value-form of Capital which now occupies a position with respect to it previously assumed by (industrial) labour in relationship to Capital, namely use-value, but the value-form of virtuality. Once stripped of its use-value of Capital, Marx's 'circuit of circulation' is the (technological) successor of Capitalism, the matrix of the will to virtuality. Ironically, just when everyone thought that Marx had been safely dispensed by historical events, in this period when the revolutionary fires of the early socialism of living labour-power have been withered away by the autocratic policies of official state socialism in the East and by official (trade) unionism in the West, at this precise moment the labyrinth of history reverses itself. The pages of *Capital* open to the section on the 'circuit of circulation' and we suddenly exit Marx's theological narrative of (living) proletariat and (dead) Capital, falling into the future of the will to virtuality in its first appearance as the pure value-form of the circuit of circulation.

For Capital, it's bitter irony. The emergence of the will to virtuality out of the economic disguise of the circuit of circulation implies not only that labour is factored out of (Capitalist) production, but Capital itself now is factored out of (virtual) reproduction. The cycle of history, with its stories replete of the rise and fall of all the dominant signs – power, war, labour, Capital, sex – breaks with the ossified structures of modernist myth, setting in motion the latest historical play, that of virtuality to be the once and always hoped for unifier of the siren of history. However, to speak of virtuality and history in one breath, to see in the circuits of virtuality a terminus ad quem of historical indeterminacy, is already to escape the present dialectic of Capital and virtuality, to say that in the valorization of the value-form of Capital lie not only the origins of the will to virtuality, but the future as the metaphysics of technicity. To say this is impossible, because to speak of the overcoming of Capital in the language of technicity, to announce Marx's famous 'circuit of circulation' for what it is, the basic code of the fully realized technological future, would be also to brush Heidegger against Marx, to speak, that is, of the hidden presence in Marx's Capital of Heidegger's will to technicity. But we all know that Marx was a prophet of the left and Heidegger a tombstone of the right, so this encircling, indeed this declension of the invisible Heidegger in the visible Marx, is not permitted. Unless, that is, the equally famous 'forgetting' in Heidegger's ontology of being is not the withdrawal of the gods from the 'desolation of the earth,' but another more proximate forgetting: the withdrawal of the visible signs of Capitalist accumulation from the (virtual) circuit of circulation. Marx, then, as the precondition of Heidegger, and Heidegger as the epiphany of Marx: a historical incompossibility.

From the 'Circuit of (Capitalist) Circulation' to the Circulation of (Digital) Circuits

In circulation, the Capitalist and worker confront each other only as the vendors of commodities, but owing to the specific, opposed nature of the commodities they sell to each other, the worker necessarily enters the process of production as a component of the use-value, the real existence, of Capital, its existence as value. And this remains true even though that relationship only constitutes itself within the process of production, and the Capitalist, who exists only as a potential purchaser of labor, becomes a real Capitalist only when the worker, who can be turned into wage-labor only through the sale of his capacity for labor, really does submit to the commands of Capital. The functions fulfilled by the Capitalist are no more than the functions of Capital – viz. The valorization of value by absorbing living labor power – executed consciously and willingly.[29]

When capitalism finally breaks with the materiality of production, entering its metaphysical stage as a pure 'value-form,' then the classic Marxian model of the Capitalist 'circuit of circulation' gives way to the circulation of circuits. While the circuit of circulation is the basis of the original Capitalist model of production, the circuit of circulation is the new model of the virtual model of production. This is a predictable outcome of the Marxian model. Everywhere in Marx's writings lurks the spectre of circulation. There is no Capitalist, only a 'Capitalist function[ing] ... as personified Capital, Capital as a person, just as the worker is no more than labour personified.'[30] Everything and everyone is a function of their relationship to the value-form of Capitalist circulation. No autonomous labour, only labour valorized as use-value; no independent Capitalist, only the bourgeoisie valorized as a command function for the extraction of surplus value from living labour power; no terminal production, only production as a twofold sign of the beginning and ending of the circuit of Capitalist circulation; and certainly no autonomous processes of consumption, exchange, and distribution, only a general circuit of circulation that requires for its growth that the (economic) phases of production, consumption, distribution, and exchange align with one another in that spectacular circular flow, that speed and magnitude of circulation, that valorization of the pure value-form of Capitalism. What was previously the fate of the individual Capitalist – his reduction to 'Capital personified' – and the destiny of the worker – 'use-value' submitting to the 'commands of Capital' – quickly becomes the ruling axiomatic of virtual capitalism. Pure circulation, pure speed, pure value-form, pure technicity.

Indeed, if Marx could speak so hauntingly in *Capital* of the relationship between labour and capital as that of 'animals and machines,' perhaps that is because his image of the circuit of capitalist circulation is deeply biological. *Capital*, then, as an extended biogenetic description of the original, governing cellular matter of the DNA of Capitalism. In this case, Marx's searing vision of workers as use-value objectified and Capitalists as Capitalism personified quickly generalizes into a biological theory of the circuit of circulation (production-exchange-distribution-consumption) as the original economic genome, not only for the historical development of the process of Capitalist production, but as the cellular material for mixing, splicing, recombining circuits of virtual clones. When not only labourers and Capitalists but also the circuit of circulation itself is violently thrown into circulation, when production is blasted apart by the globalization of manufacturing, when consumption is consumed by the circulation of signs, when exchange is digitally networked, when distribution is resequenced as a digital nervous system, then at that point we can speak of recombinant Capitalism as the destiny prefigured by *Capital*. Circuits of (virtual) consumption, circuits of (technical) production, circuits of (networked) distribution, circuits of (fast) exchange, all of which are simultaneously linked in a global model spearheaded by the circuit of virtual circulation, but all of which also immediately condense into the fourfold logic of the original circuit – production, consumption, distribution, exchange. Faithful to their cybernetic expression as the final value-form of advanced Capitalism and, consequently, the original prototype of virtual Capitalism, what matters are not the terminals of production, consumption, exchange, and distribution but the speed and magnitude of the process of circulation itself. Here, everything enters into circulation, and no one is allowed to refuse circulation; everything is circuited, no one can abandon the power of the circuit. Pure circulation then as the precondition and product of the virtual model of the circulation of circuits.

That's the real effect of the technological process of production. Not simply a circuit of (Capitalist) production, but other circuits as well: circuits of production, circuits of consumption, circuits of distribution, circuits of exchange. All linked and overlapping, all autonomous and indeterminate in their origins.

The classic model of the circuit of circulation can give rise to a process of multiple circulation of circuits because the 'value-form' can be cloned, replicated, reproduced, re-circuited. No longer embedded in the facticity of material production, circulation is about speed, magnitude, and vectors.

Nomadic Theory

Now that in the east Marxism has been severed from official state ideology and now that in the west the terror of globalization and the triumph of the culture of consumption has seemingly repressed critique everywhere, it is probably a propitious moment to return to Marx. Return to Marx, that is, not as the exponent of a hegemonic socialist ideology or as the prophet of the worst abuses of industrial Capitalism, but in the until now suppressed form of that which he has always been. Marx – as a vagabond thinker, an anti-philosopher, providing what Michel Foucault has described as 'thought from outside,' a thinker of uncomfortable truths, resisting with all the intellectual strength of a counter-will the dominant mood of the times – has nowhere to go, nowhere to be inserted, nowhere to be practised. His writings, *Capital*, the *Grundrisse*, *The Communist Manifesto*, *The Eighteenth Brumaire* lie there, abandoned detritus from a past of the nineteenth and twentieth centuries, too close to be critically examined and yet not sufficiently distanced to be fully appreciated for their possibilities and dangers, a 'thought from outside,' which simultaneously resists assimilation, and yet refuses to pass quietly away under the shroud of history.

Read Marx, then, not as a theorist of the genealogy of modern Capitalism, but as a political historian of the twenty-first century. In the same way that Marx always analysed Capitalism in terms of its 'definite historical conditions,' and in the same way that he insisted on specificity in theorizing the epochal break-point from the feudal theory of value to the labour theory of value, so too, Marx's writings resituated in the 'definite historical conditions' of the twenty-first century, which is to say, resituated in the definite contemporary context of the triumph of streamed capitalism, provide an entirely revitalized, chillingly accurate, and historically astute guide to the Capitalist axiomatic in its cybernetic form. Perhaps in one of the great ironies of intellectual life, the destiny, the fatal destiny, of Marx was not in the end to be the political theorist of nineteenth- and twentieth-century Capitalism and its communist opposition, not to be, that is, the theorist of the struggle between bourgeoisie and proletariat, but, once historically overcome in the arena of political contestation, to be finally liberated to be a really existent critical analysis of digital Capitalism. Liberated of the frozen political practices of the twentieth century, Marx is finally free to be the prophet of the Capitalist axiomatic of the twenty-first century.

The Last Temptation

And this makes sense. If you forgive Marx his frustrated millenarian ambitions to finally bring the siren-call of history to a definite close with his utopian ambitions for the proletariat as the always hoped for class of universal emancipation, if, that is, you forgive Marx for finding himself at the end of his writings, as at their beginning, in the labyrinth of history, desperately substituting the universal struggle of the proletariat for the universal spirit of rationality, if you forgive Marx for succumbing to the final temptation of trying to solve the unsolvable riddle of history, then Marx's utopian screen washes away, revealing in the vortices of political theory the glint, just detectable out of the corner of one's eye, of a strict theoretical analysis of the Capitalist axiomatic which is only apparently about the transition from the medievalism of the feudal theory of value to the modernism of the labour theory of value, but which is in actuality a mutatable and historically relevant analysis of the future of streamed capitalism. Probably against his own limited political intentions, Marx really was writing the *political history* of Capital: Capital past, Capital present, and now Capital future. In the same way that Marx transited Capitalism from feudalism to industrialism, our project is now to perform a second transition, to transit Marx from the dying order of industrial Capitalism to the emergent future of the networked capitalist axiomatic.

POLITICAL NOTES

Anti-Streaming Activism

In the same way that the early hegemony of the capitalist bourgeoisie called into existence its necessary objective political antithesis – the industrial proletariat – so, too, the swift emergence of the virtual class from the ruins of the industrial economy marks the beginning of a new cycle of political economy. Lacking a definitive name but not a definite historical presence, this 'anti-virtual' class takes to the streets and to the Net in spontaneous forms of struggle that quickly resemble a Paris Commune rebelling against the digital mode of production.

Global, connected, fungible, wireless, planetary, the anti-virtual class necessarily assumes the *technological form* of virtuality itself. But unlike the virtual class that seeks to subordinate the new mode of digital production to old relations of capitalist existence, the anti-virtual class *can only exist* on the basis of realizing the creative possibilities of the wireless world. Refusing to reduce digital reality to the language of equivalence and exchange that codes virtual capitalism, the anti-virtual class forces the virtual possibilities of digital reality to follow a *human vector*. Opposing the technocratic closure of the digital nervous system, the anti-virtual class deals in the symbolic exchange of digital dirt: noise in the system such as issues surrounding environmentalism, human rights, economic justice, labour and education. Constituted on the basis of the virtualization of knowledge, this class *of and for virtuality* represents the first militant signs of a political form of digital resistance that in its immediacy and globality is the historical successor to the socialization of labour. A Net community, a necessarily planetary community, a community of and for human rights, the anti-virtual class is, in effect, the new digital proletariat. It confronts the virtual class with a vision of the digital future that, refusing to subordinate the digital mode of production to capitalist relations of existence, challenges forth the *virtuality in digitality*. Rejecting the fetishism of money, it asks what is the *human meaning* of globalism, immediacy, connectiveness, fungibility, digital knowledge. In this struggle, the life-and-death question is: Is a post-capitalist digital future possible?

That is why the *Battle of Seattle* and ensuing demonstrations in Washington, Quebec City, Genoa, Vancouver, and Calgary with their political manifestations against the World Trade Organization were so symbolically important. Not an accidental meeting of labour activists, environmentalists, women's rights workers, deep ecologists, critical students, suburban liberals, direct action anarchists, old- and new-guard socialists, and social conservatives, but one of the first real signs of a new political class resisting the high-intensity market directives of the new economy. Perfectly symbolic, this political struggle took place in the streets of the City Electronic: a network of electronically circuited cities which with their concentration of virtual capital represent key nodes in the new electronic archipelago. Reversing the ruling logic of *faux globalization* with its abandonment of the ancient elements of earth and sky and bodies and water, the demonstrations against the World Trade Organization went directly to basic earth. Here, the politics of electronic resistance first spiked: *street politics* (not the digital nervous system); *symbolic politics* (demonstrators dressed in the beautiful costumes of soon-to-be-exterminated sea turtles, not *Bladerunner* swat squads); *media theatre* (squatting the world media, not bunkering down in closed sessions to rewrite the rules of world trade and protection of intellectual property rights at the behest of the virtual class); *direct anarchism* (disrupting even for an instant the circulation of the digital circuit in order to allow electronic space for a future of electronic perturbations). While the leaders of the World Trade Organization dealt in the market rationality of so-called 'free trade' for the virtual class, the anti-virtual class in the streets of the City Electronic put chaos theory onto the electronic screen. Not politics subordinated to the capitalist axiomatic, but a new politics that deals in the political language of chaos theory: symbolism and virtuality and perturbations and morphological changes of state, a virtual politics that exists *only because* it is aware of unintended consequences of technology and which rubs the dreams of digital reality against the *human disturbances* of third-world labour camps, environmental shut-downs, genetically engineered food, and the triumph of economic injustice.

Consequently, at first in Seattle and then in other urban nodes on the electronic vector, it became apparent that virtual capital creates two opposing moments: the first moment – the virtual class; the second moment – a planetary class of popular forces rethinking questions of ethics, politics, and culture in the technological age. Not simply a reprise of the struggle of the urban proletariat against the rising class of the bourgeoisie, the reality of globalization pits the virtual class against an emergent political opposition that has a radically alternative vision of the human situation. Here, the key political issues of the twenty-first century are joined: *being human versus the (electronic) post-human*; living labour versus a (robotic) future of post-labour; unrestrained technicity versus social ethics; an emergent concern with bio-ethics versus market-driven bio-engineering. The history of virtual capital begins with a form of electronic and street struggle between a ruling virtual class that views itself as the embodiment of a new (digital) mode of capitalist production and an emergent democratic and populist resistance that represents the *human absence* in technoculture.

In the streets of Seattle, Washington, Genoa, and Quebec City, the ideological hegemony of the virtual class was directly challenged by its own objective antithesis. If the virtual class is global in terms of its necessary control of intellectual property, then the planetary class of anti-globalization activists is necessarily universal in terms of caring for the general human interest. If the virtual class is digitally connected as a matter of economic competition, then the forces of anti-globalization come into existence on the basis of the connectivity of the general human interest. If the virtual class is digitally mediated, then protesters in the streets intensify the meaning of mediation to include the human and the non-human, the organic and non-organic. If the virtual class seeks to install a new series of economic exclusions between the digital elite and those surplus to its commercial interests, then global anti-poverty organizations are necessarily inclusive. Precisely to the extent that the virtual class seeks to constrain the possibilities of the new mode of digital production in favour of old relations of capitalist production, the struggle against globalization, this powerful expression of an emergent *human* class, seeks to make the new mode of digital production simultaneous with the emergence of a world of creative virtualities: connected but democratic, global but not exclusionary, interactive but not (technically) isolated. Beyond the economics of market rationality but corresponding perfectly with the transformation of virtual capitalism into its metaphysical stage as a value-form, the anti-virtual class introduces a debate on the *intrinsic content* of this value-form. Against the vision of pure circulation of the digital circuit, the democracy activists ask: Circulation for what and for whom? Against the twinning of capitalism and digitality, the anti-virtual class sometimes works to *intensify* digitality, to *excess* the direction and content of virtuality until the technological future is forced to disclose its *surplus-value* for a more meaningful human future. When streamed capitalism achieves its purely virtual stage as an empty value-form circulating at the speed of light, then the question of value itself is the decisive storm centre of contemporary politics. In the delirious future of virtual capitalism, there is already to be heard the ancient rhythm of the human demand for justice. Political activism is the uncompromising ethical 'no' that today marks the furthest reach of streamed capitalism.

Which explains, of course, the ferocity of official state opposition to the anti-globalization movement. Acting at the behest of the forces of streamed capitalism, one critical political function of the state is to manage public dissent, with a veneer of democratic participation always accompanied by the hard reality of policing. Since streamed media always accompany street protests against streamed capitalism, the stakes are intensely *symbolic*, and thus critical to the political stability of the government of globalization. Biotechnology firms are trumped ethically by the Brazilian Landless Peasants' Movement. The Ministers of World Finance attempting to privately work out the protocols for a new charter of transnational rights are put off balance by the demands of street democracy protesters for transparency and public accountability. The political modification of the 'politics' of the state in the direction of streamed capitalism is symbolically challenged by NGOs, anti-poverty organizations, labour activists, concerned citizens, democracy action groups, all of whom consistently and eloquently call for the return of the 'real world of democracy.' Against this symbolic challenge – democracy versus the marketplace, political transparency versus command globalization, anti-poverty versus hyper-wealth, deep ecology versus the biotech millennium – the state can, and does, react with sadistic violence: capricious tear-gassing in Quebec City, police provocation in Genoa, effectively military exclusion zones in Washington, preemptive disappearances of political activists in Toronto. From city to city, continent to continent, the symbolic protest of the anti-globalization movement is met with escalating and highly coordinated policing strategies in which state violence is brought to bear on democratic protesters newly signified as subversives now, terrorists soon.

Art and Technology

The Hyperbolic Sign of Art and Technology

What is the art of the digital matrix?

A mimetic reflection of the planetary drive to technicity or a radical form of poiesis which the universal technical state always sought to suppress and in opposition to which it imposed an increasingly hegemonic language of 'enframing' which functions now to conceal tendencies towards technological nihilism? Are the codes of technology challenged by art or has art itself become a revelation of the technological dynamo? Or is the relationship between art and technology more ambiguous? No longer coded by the language of polarities, is it possible that today art and technology reveal traces of a more ancient relationship, traces of the broken spiral of techne and poiesis the absence of which haunted the writings of Nietzsche and Heidegger and forgetfulness of which doomed Marx, against his own insights concerning the disappearance of the commodity-form into pure circulation, to the fetishism of political economy?

Against the will to technology, which is increasingly hygienic, art introduces a counter-praxis of smeared images, smudged bodies, and contaminated optics. In the artistic imagination, the 'hyperbolic sign' of Nietzsche's thought finds its quintessential aesthetic expression in streamed images of bodies moving at the speed of light, yet no less trapped in the more ancient rituals of ressentiment, scapegoating, and rage against the facticity of mortality. If the planetary drive to completed technicity is premised on what Marx theorized as the violent fetishism of the speed of circulation, then art recovers a sense of temporality – the time of duration – as a way of aesthetically undermining empty spatialization. If technoculture intensifies Nietzsche's prophecy of cynical power – a theatre of politics populated by 'blond beasts of prey' as the spearhead of the present militarization of the global economy and by 'slave morality' as popular (technical) consciousness – then art interjects itself in the form of an *aesthetics of overexposure*. As poiesis, art literally overexposes the always hidden language of power, drawing to the surface of the cynical image the labyrinth of a power which works now in the language of the despotic eye. Finally, if the project of technology is nihilistic in its essence – transforming subjectivity into an 'objectless object,' functioning by an ethics of harvesting the 'standing-reserve,' provoking boredom and anxiety as its key emotional registers – then art implicitly begins with Heidegger's eloquent admonition.

Only by listening to that which is closest at hand, only by drawing out what has been suppressed by the coming to be of technicity, can we 'think' through to another meaning of nihilism.

Putting on the wired skin of machine flesh as its privileged aesthetic medium, art today actually *wears* the speed of circulation, *breathes* the vector traces of the image matrix, *experiences* the gravitational pull of bodies caught in the web of the 'standing-reserve,' and *out-energizes* the dynamic technological drive to the harvesting of humans, nature, and animals. In artistic practice, the symbolic pull of forgotten ways of thinking – from the pre-Socratics in western culture to the mysticism of the teaching of the four ages in the east – returns as the symbolic exchange animating the wasteland of the digital matrix. Encoding itself in the digital flesh of the image matrix, art draws near to that which lies concealed in the revelation of the will to technology. It is the *aesthetic supplement* which provides tangible expression to Heidegger's longing for a poiesis that would recognize in the nihilism of 'being held out in the void' both a 'danger' and 'saving-power.' Consequently, against the deconstructive spirit of the (technical) times, art engages in a classical form of truth-telling. Its aesthetic practices issue a specific epistolary: it is only by thinking and imagining and imaging from the deepest interiority of informatics that we can create an ethical perspective that would oppose technology in the service of hyper-capitalism as the 'injurious neglect of the thing' with new ways of understanding the creative possibilities of the question of the post-human.

Consequently, understood as an expression of the life of the (digital) mind, the essays in this section – *The Image Matrix, The Digital Eye, Body and Codes* – connect deeply with the theoretical trajectory of Marx, Nietzsche, and Heidegger. Reflecting in their aesthetic concerns the technological ruins first signalled in the writings of these German prophets of the age of hyper-rationalism, these essays inject a minimal element of poetic exchange into technological discourse. In doing so, these meditations on art seek out moments of reversibility in the otherwise sealed rhetoric of technologies of the body, information and digitality.

For example, *The Image Matrix* is the aesthetic doppelgänger of Marx's fetishism of circulation. Here the virtualization of the knowledge-theory of value finds its most intensive expression in the appropriation of the archive of the global (photographic) imagination by the digital commodity-form. *Body and Codes* reverse-engineers Nietzsche. Mindful of Nietzsche's meditations on the 'maggot man' as the embodiment of two centuries of 'conscience-vivisectioning' – a body that is both a pleasure palace and torture-chamber – *Body and Codes* theorizes the twenty-first-century technologies of genetics, nano-technology, and robotics under the sign of Nietzsche. Asking what is the fate of the human in the age of post-biologics, the essays retrieves, albeit in recombinant from, Nietzsche's insistent ethical demand to understand what is happening to us – this gamble, this crossing-over, this adventure – now that we are caught up in a forced migration to the world of the post-human. The content of *Body and Codes* may be twenty-first-century, but the ethics are strictly Nietzschean. Here, *On the Genealogy of Morals* provides an ethical palimpsest for inquiring into the moral foundations of contemporary eugenic practices.

Equally, it is the spectre of Heidegger that shadows *The Digital Eye*. While Heidegger theorized the culture of boredom in the context of an intense reflection upon the will to technology of the life sciences, *The Digital Eye* surveys the ruins of the digital nerve in the context of an appeal for a politics and art of incommensurability. Juxtaposing the ancient voices of indigenous prophecy against the utopian claims of telematic society, *The Digital Eye* functions to create what Heidegger termed an 'in-flashing,' a moment of uncertainty rippling across the fabric of techno-culture. Recovering Heidegger's plea for a reconfiguration of the terms of techne and poiesis, the essay poets the contemporary language of technicity and listens intently for signs of other, now concealed, technologies of the subjugated body.

Faithful to the spirit of art which is itself incomplete and reconfigurable, these essays do not seek to install a new regime of aesthetic normativity. Quite the opposite. They are perhaps the beginning of a new conjunction between theory and art, brushing the dark futurism of Marx, Heidegger, and Nietzsche against the recombinant images of the artistic imagination. *An art and theory of techne and poiesis*. In this way, the 'objectless object' of *The Image Matrix*, the eugenic skin of *Body and Codes*, and the suppressed voices of *The Digital Eye* are creative, fragmentary, intense methods of drawing close to that which lies nearest in the thought of Marx, Nietzsche, and Heidegger: namely, understanding the will to technology as it spearheads the future of the twenty-first century.

7 The Image Matrix

Burying the Image for the Future

Today the image is so powerful that it has to be buried alive. Consider the following story:

It will be a surreal burial.

The Bettmann archive, the quirky cache of pictures that Otto Bettmann sneaked out of Nazi Germany in two steamer trunks in 1935 and then built into an enormous collection of historical importance, will be sunk 220 feet down in a limestone mine situated 60 miles northeast of Pittsburgh, where it will be far from the reach of historians. The archive, which is estimated to have as many as 17 million photographs, is a visual record of the 20th century. Since 1995 it has belonged to Corbis, the private company of Microsoft's chairman, William H. Gates.

The Bettmann archive is moving from New York City to a strange underworld. Corbis plans to rent 10,000 square feet in a mine that once belonged to U.S. Steel and now holds a vast underground city run by Iron Mountain/National Underground Storage. There Corbis will create a modern, subzero, low-humidity storage areas safe from earthquakes, hurricanes, tornadoes, vandals, nuclear blasts and the ravages of time.

But preservation by deep freeze presents a problem. The new address is strikingly inaccessible. Historians, researchers and editors accustomed to browsing through photo files will have to use Corbis's digital archive, which has only 225,000 images, less than 2 percent of the whole collection.

Some worry that the collection is being locked away in a tomb; others believe that Mr. Gates is saving a pictorial legacy that is in mortal danger...

When the move is done, Corbis's New York office will contain nothing but people and their computers, plugged into a digital archive. No photographic prints, no negatives, no rotting mess. Analog is having a burial and digital is dancing on its grave.[1]

The Death of Analogue / the Power of Analogue

The twentieth century may have been dominated by the spectacle of the image, but the twenty-first century will witness the disappearance of the image into digitality moving at the speed of light.

Not simply the death of analogue with its extended burial rights for the traditional apparatus of photography – prints, negatives, and the framing gaze of the photographer's eye – but the disappearance of the image itself. Because that is what is really at stake in this strange story of Corbis's necropolis of the photographic archive. Certainly there are serious issues of cultural politics here: issues of monopoly capitalism in digital form creating a short market in the photographic archive of the future; issues of shutting down the eye of photographic history itself; issues of substitute culture – replacing an actual worked photographic archive with its coded, and dramatically abbreviated, digital substitutes. All of this is almost self-evidently true, almost palpable in this eerie spectacle of the cryogenic deep-freezing of photography, this entombment of the reproductive rites of photography in an abandoned mine shaft in Pennsylvania. No more (photographic) images, no more decomposing smells of negatives, no more 'thumbing through' stacks of refrigerated images, no more immediacy. Now, we are suddenly living in the culture of the retrieval of digitally archived images by remote control: images safely kept at a distance from human contact, uncontaminated by the passage of time. The image archive is reduced to the steady flicker of the cybernetic code. Hygienic, sterilized, catalogued on the computer screen, untouched by the human hand, unseen by the human eye, uncontaminated by the ephemeral imagination.

But what does this really mean? Is this simply another story of the triumph of digitality over analogue – the sovereignty of the light-image over that curious mixture of light and time and chemicals that is photography? Or is this assignation of the photographic archive to the coffin of a cold underground storage vault a haunting presentiment of something more monumental, more striking for the artificiality, perhaps even naivety, of its digital illusions?

Certainly on the surface this may be a quick-time fable of 'analogue having a burial and digital dancing on its grave,' but in the strange reversals that mark the passage of life itself through the spectacle of the image, exactly the opposite may also be the case. The secret of this fable of the buried image lies in the question of the code. Because the code is what this story is really about, and it is just when we disentangle the double helix of the digital code, that twisting spiral of analogue and digital logic as they intersect and implode, that we can begin to understand the serious cultural implications of this story for the future of the image in the new century. It is in the nature of all codes, digital or otherwise, to immediately repress all signs of their opposites, to cancel from view and certainly from verbal optical articulation the repressed energies of the anti-codes that work to make possible the violence of the positive code itself. As in life so in the story of the digital code.

The digital code speaks the sanitary language of culture cleansing, of photography itself at a distance, of the archive by remote control, of the deep-freeze preservation of the image from the 'contamination' of time and history and memory and skin and smells and touch. Photography in a bubble. Memory in cold storage. Images fast-frozen. Perfectly preserved, perfectly coded. Always retrievable, always inaccessible. A psychoanalytics of digital repression.

But what if with the history of mythology as our guide, we were also to note concerning the future of the image that that which is most deeply repressed, most feared, and most preserved even to the point of its death, never fully absents itself from culture, never can be removed at a safe preserve from the future anxieties and future boredom of the enigma of life itself. In this case, it is not so much the 'burial of analogue with digital dancing on its grave,' but analogue as the repressed memory the absence of which haunts the once and future spectacle of the digital.

More than is perhaps recognizable in the orthodox media scriptures of the digital age, we are no longer living in a culture dominated by the image *because we are the pure image.* Ours is a culture signified by the triumph of virtuality, by the disappearance of the spectacle of the graven image into code. It is as if those torrents of words spilled in the decade leading up to the end of the twentieth century, those anti-words that stormed the icons of representationality, that spoke of the hyper-reality of a coming structuralist reality, finally found their moment of historical truth, not in the echoes of written language but in the language of the disappearance of the image. Hypering the image. Coding the spectacle. A hygienics of (ocular) memory. A necropolis for the photographic memory. When a culture at some deep informing cultural level finally loses faith in representationality, when it shifts its register of acceptable meanings to embrace the language of virtuality, then that culture also effaces its ability to filter memory through the apparatus of the image. The death of representationality then is also about the burial of the image, and the virtual flight from the tomb of the analogue to the new story of the cynical image.

Indeed, if the history of twentieth-century photography can be buried alive, chilled to such a degree of zero-intensity that it cannot be easily disturbed, this is simply an indicator that the image has taken flight from the medium of analogue photography to electronic imaging, from the image as a light-based product of the photographic apparatus to the vanishing of the image into the digital simulacrum. Or maybe something else. Perhaps the burial of the history of twentieth-century photography also announces in the absolutism of this gesture that the photographic image can be superfluous today because we are finally living out that age predicted by ancient prophecy – a time in which the image is made flesh.

Disappearing into Images

It was always intended to be this way.

Discontented with the radical separation of flesh and image, the body has perhaps always yearned to disappear into its own simulacrum, to become the image of itself that it thought it was only dreaming. This is why the story of the simulacra of images has nothing essential to do with the languages of domination, with the purely *social* stories of alienation or reification. Escaping from the coils of earthy mortality, the history of the image has been most seductive because of its obsessive hint of pure ocularity, because of its trance-like status as a virtual vector in an increasingly electro-optic apparatus of power. A born pervert, the image is the outlaw region of the human imagination. A natural charlatan, the image maintains the pretence that it has something to do with the history of the eye precisely because its real electro-optical history focuses on the shutting down of the eye of the flesh and the opening up of *the cynical eye of dead code*. An enigma, a sky-tracer, a going beyond, a falling back: the image is the residual trace of the human challenge to a universe that knows only the game of reversibility and seduction as responses to challenges to the power of its silence.

Consequently, it is our future to disappear into images. Not only into those external image-screens – cinema, TV, video, digital photography – but also into those image-matrices that harvest human flesh: MRI, CT scans, and thermography. The future of the media? That's the unseen cameras of automatic bank machines, the unhearing machines of automatic eye scans, the unknowing machines of planetary satellite photography. Sliced through and diced, combined and recombined, the body is an image matrix. The body desperately needs images to know itself, to measure itself, to reassure itself, to stimulate its attention, to feed its memory channels, to chart its beauty lines, to recognize its gravity flaws, age marks, and flaring eyes.

In a special case of the media preceding science, the image matrix is how biotechnology will penetrate the imagination. No need to wait for the sequencing machines of recombinant technology. The image matrix is already recombinant. No need to anticipate the results of gene sequencing: the results of the human genome image are already known.

The image matrix inhabits the body. It is the air breathed by its photographic lungs. It is the sky surrounding its digital eyes. The image matrix quick-jumps the eye and seduces the imagination. A static line. A conspiracy line. An entertaining line. The image matrix is always there.

There is no longer any difference between the body and the image matrix, except perhaps in the default sense that the body is still in the way of a falling away from the intensity of the image matrix, a gravitational pull like a dark unseen star in a distant galaxy that can only be detected by its negative gravitational presence.

Do images warp when in the presence of bodies? Like galactic star systems, do images flare outwards in the act of seducing passing bodies? Conversely, do images retract into cold sterility when animating empty spaces?

And what of light? Why is the image matrix washed out by sunlight? Is it simply a matter of physics, or something else? Is the disappearance of images when exposed to the light of the sun certain evidence that images are also possessed of the spirit of the vampire?

And what of the future of the image in the age of biotechnology?

The image is a gene machine, recombining, splicing, mutating, sequencing. No need to wait for the genetic engineering of the body because the image is already a gene sequencer, mutating and mixing culture patches.

That the history of the photographic archive of the twentieth century has now been safely interred in cold monument to the dead image only means that the final assimilation of human flesh and the image matrix is about to occur. In a culture of death, only that which has been buried is finally freed to live out the enigmatic seduction of its destiny.

A Recombinant Postscript

Saving the Future for the Image

So then, a final question: What is the fate of the image in the age of the digital? Saving the image for the future? or just the reverse: Saving the future for the image? Consequently, the urgent political question: In the digital age – Saving the image for whom?

Saving the image, therefore, for whom? and for what? The real question is not necessarily ensuring the survivability of the image, but of maintaining a cultural free and democratic accessibility to the images of the future. In effect, ensuring the survivability of an open future for the image. A digital future under the global control of the masters of the digital universe means a future of the image under the control of an acquisitive and accumulative mentality driven on by a strange, restless, but nonetheless relentless desire to possess the future of the image. Who will be the guardians of the images of the future? A Ted Turner colour-your-world future where questions of accessibility to the electronic heritage will be under the control of all the (Bill) gatesways of the world? A *closed* digital future? Or an *open* digital future? Digitally archiving images, of the future in which to access those images we will have to pass through a global networked multimedia market centralized primarily in the United States, or an open future free for creative imaging.

Not just a technical question, then, of the challenge of archiving and curating the images of the digital future, but now there is a very real cultural struggle over saving the future for creative images.

In essence, the technical question introduced by the move from electronic to digital reality might well be the implications of digital technology for the electronic heritage. For example, curating the image in a thin/client future where networked computer systems make easily possible centralized storage of the image-bank of the world's entire film history: every film, every image coded for easy retrievability, and also, of course, coded for instant digital manipulation. A digital film bank, which, if the masters of the digital universe have their way, will be much like Blockbuster Video, where a lot of independent, definitely not-mainstream films will be quickly and silently exorcised from the electronic future. A closed digital future, shut down in advance by the subordination of the Image to a digital future acting at the behest of private accumulation.

Not then so much saving the Image for the future. In the digital age, that's increasingly a transparent question. But saving the future for the image, asking the question of Images for Whom? and Images for What? is a *political question*. But it's a question which speaks of the life-and-death cultural struggle that will take place over democratic accessibility versus private intellectual property rights to the Images of the electronic future.

What's at stake is nothing less than our cultural heritage in the twenty-first century. Perhaps that is what is really at stake in these stories about the death of the image: first of the photographic image through its entombment in a new re-enactment of an Egyptian cult of the dead; and then of the electronic image as it vanishes into the spectre of virtuality.

The Despotic Image or the Bored Eye?

The digital age unleashes deeply paradoxical tendencies in the unfolding history of the image, moving simultaneously between the violent repression of the material memory of the photographic image and its recombinant recreation in the culture of the digitized imaginaire. Out of the ashes of photography under the sign of analogue suddenly appears the phoenix of the digitized image-machine. A doubled story of repression and creation?

Or something else?

If today the image proliferates with such velocity and intensity that human flesh literally struggles to become the image of its own impossible perfection –witness the psycho-ontology of cosmetic surgery – then this might also mean that we are now fully possessed by the power of the image. Not possessed by the power of the image as something somehow ulterior, and possibly alien, to human agency, but possessed by the image as a fulfilment of human desire, and perhaps desperation. In a Copernican flip, we ourselves are images to the world surrounding us: designer bodies, rip-tide abs, faces as gestures, attitudes as probes, lips like invitations, pouts like refusals, eyes like a going under. Possessed by the images once thought as somehow safely alienated as representations, *we* ourselves have become founding referents to the simulacrum that invades us.

A story of body invasion? Not really. Contemporary society is no longer the culture of the disembodied eye. Today, we play out the drama of our private existence along and within the iris of the image-machine that we once dismissed as somehow external to human ambitions. Our fate, our most singular fate, is to experience the fatal destiny of the image as both *goal and precondition* of human culture. *As goal*, the power of the image inheres in the fact that contemporary culture is driven forward by the *will to image* as its most pervasive form of nihilism. *As precondition*, we *are* possessed individuals because we are fully possessed by the enigmatic dreams of impossible images.

That we are possessed by the power of the image with such finality has the curious repercussion of driving the image-machine mad. The matrix of image-creation as it evolves from analogue to digital and now to the biogenetic struggles to keep pace with the capricious tastes and fast-bored appetites of human flesh as an image-machine. *It is the age of the bored eye*: the eye which flits from situation to situation, from scene to scene, from image to image, from ad to ad, with a restlessness and high-pitched consumptive appetite that can never really ever be fully satisfied. *The bored eye is a natural nihilist*. It knows only the pleasure of the boredom of creation as well as the boredom of abandonment. It never remains still. It is in perpetual motion. It demands novelty. It loves junk images. It turns recombinant when fed straight narratives. It has ocular appetites that demand satisfaction. But it can never be fully sated because the bored eye is the empty eye. That is its secret passion, and the source of its endless seduction.

The bored eye is the real power of the image. It takes full possession of the housing of the body. It is the nerve centre of flesh made image. It is the connective tissue between the planetary ocular strategies of the image matrix and the solitude of the human body. The bored eye is bored with its (bodily) self. That is why it is always dissatisfied. It needs to blast out of the solitude of its birthplace in the human cranium in order to ride the electronic currents of the global eye. No longer satisfied with simply observing the power of the image, the bored eye now demands *to be* the power of the image. Which is why, of course, the archival history of twentieth-century photography can now be safely interred. At dusk, the eye of the image takes flight in the restless form of the bored eye forever revolving and twisting and circulating in an image matrix of which it is both the petulant consumer and unsatisfied author.

Ironically, the bored eye has itself now become both precondition and goal for the despotic image. Which is why images can now be so powerful precisely because they are caught in a fatal miasma of powerlessness before the ocular deficit disorder of the bored eye. The despotic image may demand attention as its precondition for existence, but the bored eye is seductive because of its refusal to provide any sign of lasting interest. A love affair turned sour. With this predictable result – the increasing ressentiment of the digital image: 'Analog is having a burial and digital is dancing on its grave.'

8 The Digital Eye

An Art of Electronic Theory

There are three aesthetic codes for understanding the digital eye. Well, not really three codes, but three *anti-codes*, because I am not affiliating myself with the grand récits of technological discourse, with the digital eye as a master discourse, a gridded space, for the dynamic conquest of nature and human nature, but with that more shadowy region of otherness, of aesthetic supplementarity, those flickering, enduring aesthetic anti-codes that, like the beat-beat rhythm of the errant human heart or like the tearing up of the eye when it gets some digital dirt in it, just won't go away.

I haven't got any modernist illusions: I know the rules. I have studied my Heidegger. I sense how deeply Heidegger was correct when he said that in the twisting spirals of the dialectic of life and art and technology and the human spirit, otherness, even dialectical otherness, aesthetic anti-codes, work only to confirm the existence of the codes, and that, ironically, anti-codes probably serve one last great normalizing function – they work to energize the code, to establish its limits, but also to strengthen its resistance. Jean Baudrillard has been this way. He called it 'the terrorism of the code' in the logistics of production. So has Octavio Paz when he said of de Sade:

> Prisoner in your castle of crystal of rock
> You pass through dungeons, chambers and galleries
> Enormous courtyards where the still black poplars dance.

Note: Several stories in the section focusing on hyper-perception were written in collaboration with Marilouise Kroker.

All is mirror!
Your image persecutes you.

Man is inhabited by silence and by space.
How can this hunger be met and satisfied?
How can you still the silence? How can the void be
Peopled?
How can my image ever be escaped?[1]

Roland Barthes went to his death still puzzled to the point of intellectual and aesthetic paralysis by the shock of trying to understand the full
implications of this game of code and anti-code. He called it the empire
of the sign, the empire of the cynical sign, and as he regressed in books
such as *Barthes by Barthes*,[2] into those picture albums of memory, of
photography as always tinged by the solace of death, he muttered to
himself words like: 'I am a public square;' 'I am anachronic subjectivity;' I am "degree zero."' The same thing today for Paul Virilio, who in
the first chapter of *Open Sky* comes to an insight on the truly radical
implications of the quantum physics and relativity theory of the 1930s
for understanding hyper-modern culture, that digital technology actually moving at the speed of light has finally overcome the resistances of
local time and local space, blasting us into the light-time and light-
space of 'eyeball culture,' that we're plunging upwards, falling upwards, into the dimensionless space and vectored time of the new
world city of 'electrooptics.'[3] In every one of his books, *War and Cinema*,
The Vision Machine, *The Art of the Motor*, *Speed and Politics*, *Pure War*,
Virilio is haunted, like Baudrillard, Barthes, and Octavio Paz before
him, by the *electronics of perception*, panicked into getting off the freeway
of accelerated culture, with its dromological vehicles pirating human
flesh in the name of cyber-sexuality and telepresencing, getting off at
the double exits marked traditional religious ethics and traditional
sexuality.

  But why panic at accelerated culture? Why not do the reverse? If, as
Marshall McLuhan always argued, artistic vision is a point of 'maximal
sensitivity' to the blast of technological change, then why not insist that
the method for understanding emergent technologies, the theoretical
practice of a new electronics of perception, be deeply infiltrated
and contaminated, indeed, overtaken and over-coded, by the hyper-
aesthetics of electronic artists. Not a theory of electronic art, but an art
of electronic theory, then, for understanding the visual codes of the

actual electronics of perception in the light-time and light-space of culture under the optical sign of the digital eye.

Why not?

Implode the code.
An aesthetics of hyper-perception
An art of excess for a time of moderation.
An art of interminability for a culture fast-cycled with changing fashions.
An electronic art that rubs together code and anti-code, and refuses the designation of the polarities.
Art like bubbling electronic asphalt on a hot summer day.

So, then three anti-codes for an art of electronic theory:
Its ontology: digital dirt.
Its political focus: technologies of otherness in the cybernetics of the everyday.
Its aesthetics: digital incommensurability in the age of calm technology.

The 1st Anti-Code: Digital Dirt

What's the smell of blood on the digital tracks? What's the sound of static deep in the wires? What's the colour of electronic discharges as they bond flesh to the machine? What's the speed of the body when it has been force-fed by high tech? What's the rate of burn of the digital nerve as it blasts off from the gravity field of the human sensorium?

The ruling illusion of digital reality is its antiseptic cleanliness. A virtual hygiene movement that launches a global tech style that is clean, cool, and cold. Virtuality as about the digital scrubbing of the world.

The will to virtual hygiene can be so powerful because the really existent animating force of digital life is dirt. Noise in the machine. Liquid in the wires. Waste in the System. Accidents in the codes. Distortions in the gifs. Mutations in design. Data Crash. Indeed, it is the absence of dirt that haunts the virtual hygiene machine and without which the system as a whole loses energy, running down to digital entropy. Ironically, the virtual hygiene machine requires for its reproduction the anti-code of digital dirt.

Scratch digital, perverse robots, spew electronic music, splatter cinema, dance of mutant life-forms, artificial perversity, spit writing, fleshmatic theory, multimedia graphics that nitro-burn ram memory – that's digital dirt. A raw, unfinished, visceral level of life-giving energy that animates the digital, seducing the codes of the digital eye from going to ground in the sterilities of the virtual hygiene movement. So then,

Excess everything.
Intensify everything.
An end to all mediation by the codes of visuality
Or maybe just the opposite
Struggle to begin again and again
the over-saturation everywhere
of the codes of augmented visuality.
An art of symbolic exchange
An art of hyper-perception
Storming broken interfaces,
Searching out lines of electronic fatigue,
Occupying digital slash marks between the either and the or
An electronic wound that refuses to heal
An art of hyper-perception
For a culture that is dying because of its lack of symbolic renewal,
Of its lack of mythic renewal

The 2nd Anti-Code: The Cybernetics of the Everyday

Digital dirt creates counter-visions to the digital code by privileging the cybernetics of the everyday. Not just the everyday present and future, but the everyday past as well – listening also to ancient prophetic voices for what they have to tell us about 'other' technologies of otherness.

In her book *The Eight Technologies of Otherness*, Sue Golding theorizes another way of being technology.[4] Not technology as a violent field of instrumental reason, but more like something Foucault might have thought in his Death Valley days, technology as techne, as a way of doing things, as a way of escaping the shroud of identity politics with its over-privileging of binary logic, and speaking not of difference, but of the indifferend – technologies of otherness – technologies of dwelling, noise, contamination, cruelty, curiosity, appetite, skin, nomadism.

As Golding says: 'What if we were to stop sterilizing the wounds? What if we were admit that the so-called deep and violent cut of meaning, truth, death, identity itself – the "who we are" and "what are we to become" of science and of life – have collapsed under their own bloodless and sexless weight.'[5] Technologies of otherness, then, dwelling, comtaminating, moving, nosing, being curious, being cruel, being multiple in the valley of the cybernetics of the everyday.

So why not too – *technologies of disappearance* – as a way of keeping alive deep memory and deep time? A grand unified theory of prophecy in the bedlam of a coarse electronic culture? Nobody is listening, except maybe for the wandering spirits of the disappeared and, of course, the day and the night and the moon and the eagle and the trickster raven and tired, really tired, human flesh. Animism is nothing to be ashamed of. It's repressed knowledge. It's the forbidden eye. It's the ticket to the truth of prophecy, of mythic utterance. That's why it has been so discredited by all the demon-spirits of the virtualizers. An art of electronic theory also begins with digital animism – that's what we've really been deprived of in the 'intimations of deprival' of the technological dynamo.

Like recently, at that epochal event, at the gates of Yellowstone National Park, where another technology of otherness took place, another cybernetics of the everyday, another technology of disappearance. Suddenly early in the morning a hundred Plains Indians gathered for the first public performance in a hundred years of a dance honouring the buffalo. Now, the urgency was directly political – to change the present policy of the Park Service, instigated at the behest of western ranchers, where buffalo who haven't it seems yet learned anything at all about park behaviour and park boundaries and being-buffalo for high-velocity tourism, can be shot, and are shot, by the thousands if they stray outside the buffalo bunker archaeology of the park. Now, the reason is political, but the overwhelming feeling is mythological. The Plains Indians acknowledge deeply the truth of ancient aboriginal prophecies that the final death of the buffalo means the final death of the people, but also that the return of the buffalo anticipates the return of the people.

The report was as follows:

As a few hundred people watched beneath the snow-dappled peaks on the northern edge of the park, Gary Silk, a Lakota Sioux, stripped to the waist. As he stood in the numbing wind, in the shadow of the giant stone arch that greets visitors here, other members of his tribe made two incisions in his back with a surgical scalpel, inserted a stick in each wound, and tied a buffalo skull to each stick with ropes.

Then, as 100 or so Indians sang and drummed and played an eagle bone flute, Mr. Silk danced around a large circle dragging the skulls through the mud and grass behind him.

After a half-hour of dancing, Mr. Silk stopped and someone sat on the skulls. Mr. Silk grabbed a horse's tail and the animal pulled away, ripping the sticks from his bloody wounds. Then, as people wept, Silk joined the singing and the drumming.[6]

Now, what do slaughtered buffalo and the Lakota Sioux have to do with the digital eye? It turns out: maybe a lot. The digital eye, like the eye of the hunters of the western lands before it, sees nothing but its own restless movement. The Lakota Sioux look into the eye of the buffalo and see their past, and maybe our future.

The cybernetics of the everyday, then, as also about recuperating the language of animism and mythology and prophecy for an art of electronic perception. Its admonitions: Honour the disappeared, Respect futility. Counter the war spirit of electrooptics – 'eyeball culture' – with the voices of prophets with their sticks and scalpeled backs and crushed skulls and blood sacrifice. Today, it's animism most of all that takes flight in the gathering of the digital dusk.

The 3rd Anti-Code: Digital Incommensurability in the Age of Calm Technology

There's a new vision of the digital future coming out of the research labs at Xerox Parc in Palo Alto. It's called 'Calm Technology.'[7] According to Mark Weiser, Xerox's chief technologist at Parc, computing has had three main trends – mainframe computing where many people share a computer; the personal computer typified by a one-to-one relationship between the PC and its user; and now, facilitated by the possibilities for distributed computing on the Net and the Web, the next new digital media will be 'ubiquitous computing' and 'relational processing,' where, as Mark Weiser says: *'many computers will share each of us.'*[8] The tech hype spins out like this:

The third wave of computing is that of ubiquitous computing, whose cross-over point with personal computing will be around 2005–2020. The UC era will have lots of computers sharing each of us. Some of the computers will be the hundreds we may access in the course of a few minutes of Internet browsing. Others will be imbedded in walls, chairs, clothing, light switches, cars, in bodies – in everything. UC is fundamentally characterized by the connection of things in the world with computation. This will take place at many scales, including the microscopic.

There is much talk today about 'thin clients,' meaning lightweight Internet access devices costing only a few hundred dollars. But UC will see the creation of thin servers, costing only tens of dollars or less, that put a full Internet server into every household appliance and piece of office equipment. The next generation Internet protocol can address more than a thousand devices for every atom on the earth's surface. We will need them all.

The social impact of imbedded computers may be analogous to two other technologies that have become ubiquitous. The first is writing. The second is electricity, which surges invisibly through the walls of every home, office and car.[9]

With embedded processors and the Web as harbingers, ubiquitous computing will light up the digital future.

However, what will be the human response when the realization grows that *'many computers share each of us'* and that we've become exactly what McLuhan predicted – the wired sex organs of the cybermachine allowing it to fecundate and develop while it reciprocates with increases in personal wealth'; 'passive servomechanisms' of an externalized central nervous system; or what Virilio foresaw – electrooptics as a parasite/predator boarding the metabolic vehicle of human flesh, from without and now from within, interfacing every orifice of the human sensorium with artificial plug-ins, with such intensity that the artificial environment that is really being monitored and managed is data flesh itself? In ubiquitous computing, we become figure to the ground of technology, body environments to the electronic sensorium. The flip is complete. At Xerox Parc, *bodies are digital interfaces.*

Thus, the urgent necessity at Xerox Parc for 'calm technology,' for making the final interfaces of flesh and machine not frenzied, but 'calm and comfortable,'[10] not a scratch in the digital eye that continues to hurt, but a 'calm technology' that attaches itself directly to human perception, 'engaging both the center and periphery of our attention, and in fact moving back and forth between the two.'[11] As a logistics, tactics, and strategy for derealizing human perception in the age of ubiquitous computing, calm technology has three aims. The first is to move most actual computing interfaces to the periphery of human attention. As researchers at Parc like to say,

> Things in the periphery are attuned to by the large portion of our Brains devoted to peripheral (sensory) processing. Thus the periphery in inform-ing without overburdening. Second, by recentering something in the periphery we take control of it (or at least have the illusion of control).[12]

And finally, by encalming human perception in this new ubiquitous experience of being a transmission vector in an array of cybernetic data, drowning in the cyber-system, calm technology 'puts us at home, in a familiar place.'[13] Calm technology provides an 'information visualiza-tion technique' for home-grown 'locatedness' in the digital world.

Now, I have chosen to bring Xerox Parc's theory of calm technology from the periphery of digital futures research to the centre of attention because it is a premonitory sign of the actual electronic war strategy by which the digital eye seductively, but no less violently, will colonize human perception. *Today, the eye is the key interface.* Not only coloniza-tion of the objects of perception without, but an endocolonization of perception where the epistemology of perception is broken into and digitally reconstituted by all the Xeroxes of the digital eye – questions of the meaning of 'locatedness,' issues of what constitutes periphery and centre, what is cyber-figure and cyber-ground in the inverted rela-tionship of the human sensorium and ubiquitous computing, ocular strategies concerning how to transit perceptual focus between centre and focus using newly acquired digital information visualization tech-niques, cyber-psychoanalytic theories that privilege this strange, opaque, diffuse ocular state of 'calm perception' for an era of 'calm technology.' Calm technology is one of the master visual codes for growing a new digital eye for the biotech millennium. Prozacing human perception.

What is a *counter-epistemology* to the calming of human perception in ubiquitous computing?

In his evocative meditations on the virtual art of Marcel Duchamp, specifically on Duchamp's *Glass*, where two mirrors held at right angles capture in the infinity of their impossible optical regression a foreshad-owing of our own refraction into the dissimulative space of virtual reality, Jean-François Lyotard spoke of our existence today as a 'hinged experience':[14] an electrooptical universe that is multiple, incongruent, and fully incommensurable, a universe that is all a matter of 'strange projections,' 'vanishing gateways,' 'partition walls,' 'anamorphoses' – a hinged universe. Now Lyotard always viewed artistic vision as a 'transformation matrix' – foreshadowing a new politics of incommensurability. Or, as he said:

> But the discovery of incongruences and incommensurabilities, if one brings it back from the space of the geometrist to that of the citizen, obliges us to reconsider the most unconscious axioms of political thought and practice. If you haven't despaired of your life on the pretext that all justice is lost when incommensurability was lost, if you haven't gone running to hide your ignoble distress beneath the nobility of a great signifier capable of restoring this geometry, if, on the contrary, you think like myself that it's the right moment to render this geometry totally invalid, to hasten its decay and to invent a topological justice, well then, you've already discovered what a Philistine could be doing searching among the little notes and improvisations of Duchamp: materials, tools, and weapons for a politics of incommensurables.[15]

Now, many intellectual nomads have been along this way: David Kristian's tattooed sound; Steve Gibson's transgenic images of *Telebody*; Tim Murray with his theory of 'digital incomposibility,' Kathy Acker with her writing as a 'transformation matrix' in *Pussy: King of Pirates*, the new school of hyper-raw edge mechanix poetry in the Bay area – visual incongruities, visual disturbances, – all hyper-modern, all posing the same aesthetic gesture of a politics of incommensurability as a way of refusing to restore the geometry of normalized reason and normalized sight.

I emphasize Lyotard's plea for a politics of incommensurability in the context of Xerox Parc's proposal for the calming of human perception at the advent of calm technology, because we are suddenly in the midst of Lyotard's ocular world that we thought we were only theorizing, except this time the visual stakes have been aesthetically raised. As if by a tactic of a preventive first-strike optical inoculation, the age of ubiquitous computing, with many computers sharing each of us, suddenly softwires human perception into an actually existent electronic universe of 'strange projections,' 'vanishing gateways,' 'partition walls,' 'anamorphoses' – a hinged universe at the interface of flesh and machine – *only to immediately shut down both the danger and the creativity of this new ocular region by the electrooptical ideology of calm technology.*

The aesthetics of Digital Dirt finds its real political nemesis in calm technology. Calm technology simultaneously closes the eye of human perception and opens human flesh to full passive absorption into the regime of the digital eye. I follow Lyotard: No to the new grand signifier, No to the new grand ocular policing, No to calm perception for a calm digital eye; and Yes to Lyotard's thought that 'it's the right moment to render this geometry totally invalid, to hasten its decay and to invent a topological justice.' However, unlike Lyotard and Duchamp, for us this moment, this right moment, is more difficult because today we do not have to simply highlight missed chances for incongruities and incommensurables, *but actually have to seize back from the optical regime of the digital eye the language of art and technology as a 'transformation matrix,'* replete with strange projections, electronic voids, virtual gateways, digital anamorphoses – a 'hinged universe' of impossible perspectives and 'topological justice' that desperately requires that its digital future, its ocular future, its electrooptical future, be opened up – *excited* – rather than shut down and calmed.

So then, some stories of excited perception – of hyper-perception.

An Eye That Hears

What if biotech does something fundamentally different from previous technologies, not simply disturbing the ratio of the senses, but reconfiguring the senses, creating mutations and hybridities of the previously separated ratio of the senses? Not just the human ear acoustically augmented to the enhanced hearing of a cat or the nose with the smelling power of a sniffing dog, but the human eye as a digital eye scanning the sky like a migrating bird, washing out the sun and the moon in favour of suddenly visible lines of magnetic polarities across the horizon or the human nervous system jacked up to such a point of receptivity that, like cattle before it, the body's nerve net sets off early warning systems for approaching earthquakes.

An eye that hears. Skin that speaks. An ear that sees with upgraded 20/20 vision. A recombinant body with tactile smell, touch that arcs across the colour spectrum, chromocratic sounds, muffled sweat, talking retinas and noise that bleeds, Lasix eyes, eyes that see but have no vision.

Of Mermaids and Werewolves

To speak of a future of hybrid organs of perception is, of course, to return to a deeper anxiety in the human spirit, an 'anxious object' expressed most eloquently and hauntingly in tales from classical antiquity of other hybridities, other mutants – of mermaids and werewolves and minotaurs and centaurs, of mutant beings half-man/half-bull, half-flesh/half-machine.

But why the anxiety over mutant flesh? Perhaps because of a more primitive fear of the inevitable mythic punishment that follows this challenge to the gods for such a radical transgression of the language of species difference. Icarus may have been doomed not for his hubris in seeking to escape the gravity of the earth's downward pressure on human skin and bone and blood, but for his revolt of the restless heart on behalf of a new species being: part-bird/part-human. Transgressing species difference is the limit experience of classical mythology – the in-between, the third interval, the eye that turns upward into the white soft mass and hovering dark orbit of the skull's cavity. It is to be feared with a frenzy of anxiety, and desired, secretly desired, but desired nonetheless with the passion of finally transgressing the forbidden limit.

But what if the mythic fear of mutating boundaries is also to be understood psychoanalytically for what *is*, has also always been? A strange movement of desire to be the forbidden limit and beyond: being wolf, being cat, being strange, being hybrid, being cloner, being splice, being fish swarm and animal pack and cougars in the soaring mountains of the electronic fields, being sun skin and photosynthetic eyes and G4 memory and flash talk and streaming vision. An iMac personality with a vrml smile and a tasteless tongue cut with lynx ears and a feline software profile turned to the speed of light.

Perhaps the mythic past with its fables heroic and tragic and ambivalent of the meeting of gods and man and history is our recombinant future. A Digital Achilles. A Flash Art Ulysses. A Homer for all the technological fundamentalists. After the long sleep of a radically divided and radically rationalist modernism which repressed the eternal validity of myth with cynicism and self-hate, we can hear again the first intimations of other travellers at future's gate. Minotaurs and werewolves and centaurs and mermaids and ravens and eagles begin to stir again from the sleep of repression, stories forbidden of the Taquesta Indians at the mouth of the Miami River begin to circulate again with their haunting insight that the soul is to be found in the shadow, the reflection, and the eye, the Haida story from the Queen Charlotte Islands of the clam shell as the beginning of the life of the people, Hopi prophecies of an age that is doomed without the

sheltering sky of mythology, the modern ratio of the senses that refuses its divisions and equilibriums, and goes directly for the language of transgression – transgressing the ratio of the senses with
strange hybridities – disturbing the eye, cloning the ear, shock-
waving the real.

The Biotech Eye

The eye of the future is the biotech eye
The matrix machine
The mirror moon becoming trash
A slow eye in a fast zone – a data catcher.
The eye of forgetting – the sun shattered eye –
slips through the code and captures light.

The eugenic eye opens up into a universe of artificial flesh, chip nerves,
laser lips, cloned organs, mutant DNA, vector images, and virtual
dreams.

The wired universe grows an eye, and we become travellers of the
digital iris.

Soul Catcher

A mind machine code named the 'soul catcher' is being developed by
British Telecom. In the future the soul catcher will be implanted into
the brain to 'complement' human computational skills. The soul
catcher would enable the gathering of extrasensory data by the human
brain – in this case, data transmitted directly by wireless networks.

Scientists predict that once in place, the soul catcher will serve a
double function. Not only acting as a superconductive medium
linking the human brain with neural networks of data, but also an
internal surveillance function – signalling chip manufacturers from its
privileged position within the human brain about the human central
nervous system. A researcher at British Telecom stated that the 'future
of the human species depends on our continuing ability to process
information.'[16]

So, you become the machine.
The machine becomes you.
The brain of the Soul Catcher
As a neural network skull
Fast Processing
Fast Data
Waiting finally to see with its biotech eye.

The matrix machine. The mirror moon becoming trash. A slow eye in a fast zone – a data catcher. The eye of forgetting – the sun shattered eye – slips through the code and captures light.

Tongue Talk

The Tasteless Tongue

Imagine the eye
But taste the
tasteless tongue
Empty in the mouth
Opening wide and
Speaking in waves
Of hybrid sounds
No bitter, No sweet
To dilute
The no-taste taste

An artificial tongue has been designed by researchers in Texas. Eventually the virtual tastebud tongue will be able to distinguish bitter, sour, salty, sweet, and all possible variations. According to the Texas engineers, the electronic tongue relies on compounds that act as sensors, changing colour in the presence of certain chemicals. The Texas scientists found a way to attach the sensors to tiny beads that are then placed in microscopic cavities machined into a silicon chip. Tastebuds on a chip.

An artificial tongue, then, for an artificial universe.

Perhaps not just an electronic tongue for the food and beverage industries to taste their products, but the artificial tongue as the first biotech sign of our deep immersion in a new media universe of tongue talk.

Tongue talk for a universe of artificial taste, where the tongue grows the electronic noosphere for lips, rips itself free from the liquid prison-house of the mouth, abandons the painful history of speech and slurs and stutters and spit and silences, and grows a universe, a tongue universe, for a horizon of the tasteless tongue. The data tongue.

When the data tongue begins to speak it's always in the baby talk of the binary code. But the data tongue doesn't stay a baby for long. It licks flesh. It licks genes. It's curious. Tastes good. Tastes bad. Tastes no-taste. The tasteless tongue. Da Da binary soon goes Da Da. The tasteless tongue seeks out the borderlands. The tongue spits out the data machine.

It smacks its lips recombinant with a slurping sound. Tastes have codes. Tastes have colours. Tastes have sensors. Algorithmic codes for a tasteless tongue suddenly gone aesthetic. The tasteless tongue is born with a mission to perform. It's created as a taste machine in the service of a virtual capitalism without imagination. Code Taste: streamed data where the universe begins to taste itself, to lick itself, to kiss and suck and colour and code its tastes recombinant. The planet grows wet lips of cold data and an artificial tongue of hot codes.

The chrome tongue. The silicon tongue. The chip tongue. The mirror tongue. The vision tongue. The tongue with distended virtual eyes for better licking the digital future.

The Genetic Matrix

Today it is almost as if the brilliant sun of digital reality has hidden from view the real technological future of the twenty-first century: the interfacing of advanced cybernetics and radical experiments in biotech to produce a millennium of biotech flesh.

Consider these examples:

Disabled Embryos
Researchers at a small biotech company in Boston want to produce a disabled human embryo – fatally flawed but capable of living for several weeks – just long enough for scientists to harvest the primitive stem cells so coveted for brain and heart research. The president of the biotech company engaged in the research had this to say about the production of disabled human embryos: 'The trick will be to do just

enough damage to preclude the cells developing into a viable being, but not so much to interfere with the growth of stem cells for harvest.'[17]

Molecular Breeding
The California Institute of Technology is developing a new science of artificial or 'directed' evolution. According to a recent report:

> The idea is to imitate the natural processes described by the biologist Charles Darwin in *The Origin of Species* 140 years ago. First comes muta-tion to create variations, then selection of the best variants. Then, breeding or sexual recombination to produce a new, improved generation.[18]

Now, a swifter pace of evolution is necessary for pharmaceutical com-panies, where the first company to market a drug enjoys enormous financial advantages. And also keeps up with its enemies – bacteria and viruses – that develop resistance to antibiotics by rapidly mutating themselves. The solution is 'molecular breeding': 'Imitating natural evolution, scientists grab enzyme genes out of a cell, break them up, and zap them with chemicals to alter their DNA. The result, after 5 generations, new strains of enzymes appear':[19] *artificial evolution – sex in a test tube, genes in a jar* – with unlimited possibilities for cloning, mutation, and recombination. As one scientist said: 'We're mimicking natural evolution over hundreds of millions of years.'[20] Artificial evolu-tion at light speed overpowering the resistance of natural evolution. Quick-time evolution with no need for mutation, creative variation, breeding or sexual recombination to improve the genetically engineered generations of the future.

Light-Through Flesh
In Japan, biotech companies have already genetically engineered mam-mals lit from within – *phosphorescent skin*.

Jurassic Park Recombinant
Advanced Cell Technology, a Boston biogenetics firm, has attempted to preserve an endangered species – the guar – in the womb of a cow in a field in Iowa. Not so much a futurist scenario of Jurassic Parks for extinct species, but something more ominous. As a microbiologist re-cently reported, the CEO of Advanced Cell Technology is obsessed with the failure of science to overcome the problem of human mortality. Remembrance of Francis Bacon's *Novum Organum*. Perhaps the guar is

a momentary substitute for the future retrobasing in a petri dish of the human species. That the baby guar died of dysentery two days after birth was not viewed as a terminal setback, but only as a necessary casualty on the way to a future of artificial life. As Nietzsche murmured: the ressentiment of reason grows directly of the failure of the body to overcome 'time's it was' – the inevitability of the natural cycle of life and death.

New Media Rabbits
In France, the new media artist Eduardo Kac has recently collaborated with French geneticists to transplant the fluorescent genes of a jellyfish into the body of a rabbit. *Artistic vivisectioning.*

Mutant Fish
In Finland, it was recently reported that genetic engineers have developed a prototype for farmed salmon potentially weighing 550 pounds: a futurist vision of transgenic animals, fish, plants, and humans. *A transgenic future of farmed flesh.*

Sequenced Skin

And why not? It's the age of the streamed body, the data body.
    The body has always been the site of the most radical political, indeed eschatological, struggle – the decisive site for the inscription in flesh of power as its speaks the body future.
    The data body is the recombinant body: cloned by the biotech industry, spliced by artificial skin, digital nerves, and networked intelligence, resequenced by the liquid signs of brand-name consumer advertising. Simultaneously the targeted axis of the interfacing of digital reality and biotechnology and the site of future political struggle where flesh rubs against the will to virtuality, the data body is, for better and for worse, the spearhead of techno-culture.
    In the age of Christianity, the body was virtualized, split into warring bodies of flesh and grace, with the corporeal body undergoing almost two millennia of dogmatic purification, with fire, with the rack, with pincers, with the rope. In the age of capitalism, the body was commodified, sometimes colonized as exchange-value, invested by all the signs of advertising culture in that fateful transition of capitalism from the commodity-form to the sign-form, vivisected by

a fourfold strategy of domination, from alienation (Marx) and reification (Lukács) to simulation (Baudrillard) and now virtualization. In the age of technology, the so-called autonomous body, this always doubled body of flesh and grace and use-value and exchange-value, shatters into a thousand digital mirrors. The data body. The android body. The mutant body. The designer body. The cloner body. The transsexual body. Digital flesh loop-cycling furiously within the limited space and time of a single (biological) life cycle: indeterminate, neutralized, floating. The data body itself as *the* new media future.

With or without our consent or public discussion, the digital future leaps beyond the old forms of twentieth-century politics, finance, culture, and society to create an unpredictable future in which the programmer, the engineer, the eugenicist, the multinational multimedia czar install the ruling codes of the digital eye.

What is the future of digital reality? What is the consequence of the fateful meeting of digital experience and biotech engineering? What is the fate of the future itself when, as Paul Virilio argues in *Open Sky*, time and space as the deep horizon of our existence have been accidented, have been radically derealized, into the dimensionless void of 'space-light' and the real time, the instantaneous, global networked time of light-time? After the mutation, the galactic debris of local time and local space gets in our eyes, and history as a chronological succession of events collapses into random events with mutable meanings. Or when reality suddenly flips, and we are no longer living in hyper-reality, broken boundaries moving at the speed of light, but just the opposite – *reversed into a digital universe moving at the slow speed of light*.

The Slow Speed of Light

Just in time for the twenty-first century, physicists have frozen, and then actually *stopped*, a beam of light, thus instantly reversing one of the threshold laws of physics and more importantly reversing the hyper-speed of the culture of acceleration into slow optics, slow media, slow light.

When light moves at 38 mph, decelerating from over 186,000 miles per second, and then is suddenly frozen in its trajectory, hyper-reality crashes under the accumulated weight of deceleration.

Crash optics for a time when light-time and light-space decelerate into a sub-time and a sub-space where human perspective is suddenly faster than the electronic transmission of the image.

Crash time when the slow speed of light, the slow speed of electronic perspective, brings down to earth all those globalized vectors of simulated time that had been launched into global media orbit during the reign of the speed of light.

Crash space when the slow speed of light deflates all the trends to globalization and real time into the slow time and the slow space of the frozen local.

This sudden and catastrophic deflation of the speed of light amplifies the crisis of the real. The hyper-real was based on the accidenting of bounded light and bounded space with the disappearance of light and time into the 'real time' of virtual light and virtual space. An aesthetic regime of signification based on the speed of light. What happens now is the end of the hyperreal, and the beginning of the sovereignty of the subreal. Light moving at the speed of a car stalled in gridlock on a LA freeway in the noon sun. Light that moves slower than the propagation of electronic images, and the circulation of electronic sounds.

A slow speed of light for a culture about to undergo a fast descent into a vertiginous experience of pure virtuality: events spinning outwards faster than their images; slow light as a form of retinal persistence of the slow unfolding of past events; life imprisoned in slow time and slow space and slow images and slow aesthetics. Interminability as the dominant cultural sign of the twenty-first century.

Firewire Eyes

Heidegger was correct. Everything to this point has been preparatory, an anticipation of a fundamental and radical technological event. A decisive *turning*. Until now, the will to technology has been on the outside, hovering around the body with its probes of seduction. It invites the body to empty itself into wireless networks. It firewires the eye to images recombinant. It harvests flesh. It whispers about the speed of data flesh. It undermines the confidence of the body, making it increasingly insecure about itself. It reconfigures the brain. It externalizes, ablates, and disappears flesh: turning the organic body inside out into an open-source scan portal for probing media of communication. Until now, everything has been about the *exteriorization of the central nervous system.*

Not for much longer, though. The real implication of biological determinism as the dominant discourse of the twenty-first century – the human genome project, nanotechnology, therapeutic cloning, genetic engineering, the convergence of artificial intelligence and robotics, the dreams delirious of designer genes, organ transplants and tissue replacements for an improved posthuman sensorium – is that *the body is about to pirated by the genetic matrix.*

Pirated not once, but twice. First, biotech invades the body in the name of good health, longer life, better learning, avoiding catastrophic illnesses. It faciliates the disappearance of the body in the name of its own improvement. It provides an automatic internal surveillance system of the previously autonomous body from within its own circulatory systems. And then, biotech pirates the body away from itself in the name of a perfect eugenics. Literally, the organic body is about to be replaced, redesigned, and left behind as gene kill by biotechnology acting as a predatory war machine. Read the business pages of any newspaper. The final harvesting of data flesh is the newest IPO. This future is not for everyone. Most will be left behind as surplus flesh. The genetic elite will pass over. Visions of *Gattica.* Their children already have. They will be the leaders of the new world order of the genetic matrix.

Heidegger said that to understand what is closest at hand sometimes we have to travel furthest. So read the hermetic manuscripts of the desert mothers and the desert fathers. And not just the ancients, the Anchorites, but the desert thinkers of the biotech future: digital artists. The visionaries of the biotech eye who have gone ahead by travelling furthest to what is nearest to us: our bodies. Long dead words blowing in the sand of caves in the deserts of North Africa or in other deserted caves in the deserts of the digital vortex.

That which is coming has long been predicted.
That which will happen has long been prophesied.
That which will disappear has long been lamented.
That which will be dominant has long been feared.

The twenty-first century is an age of tremulous technological destiny. The future is suspended in an electric shadowland of ambivalence, drifting between fascination and dread. Both tendencies necessary. Both tendencies incomplete without their opposite. In biotechnology, a new form of android life – the final revelation of technology as a new form of android life, a posthuman life substituting itself for the human species, finally begins.

We are one of the last generations before something fundamentally new. The experiments have already begun.

TRANSGENIC ART

Art, today, rides the digital nerve to the genetic matrix. Abandoning the wired flesh of the twentieth-century body, the art of the new century quickly reconfigures itself into the aesthetic codes of the post-human. Transgenic art for an increasingly transgenic culture under the sign of biotech. Here, the sounds and images of genetic engineering – cloning, resequencing, tissue replacement, organ farming, spliced flesh, coded genes – are ejected from the hermeneutics of biological scripture into recombinant art for a culture of biotechnology. In transgenic art, the controlling codes of genetic determinism finally flee the skin of the body, exhibiting their hyper-aesthetic possibilities for mutation, for resequencing, for cloning, for regenerative medicine in the language of the screen, in the images of sound. And transgenic artists? Not so much probes of an unknown future of technicity as aesthetic registers of the fatal destiny of the gene. Transgenic art, then, as violent, seductive probes of the body by the genetic code.

The Transgenic Art of *Telebody*

Steve Gibson's new media performance, *Telebody* (www.Telebody.ws), is a mythic story of the code seductions of the transgenic body, the transgendered body. An art of doubled sex.

A multimedia techno-sound composer by profession, a recombinant performer by choice, and a third-sex body by desire, Gibson has projected into the eerie floating images and sounds of *Telebody* all of his own speed desires to let his flesh dissolve into the eddies and psychic whirlpools of all the string theories and tunnelling effects and warp jumps of the new media galaxy. *Telebody* is not so much new media performance art as a sonic blast off into the gene-time and

gene-space of recombinant culture. In its high-distort noise and liquid
images the castle of modern referentials collapses under the pressure
of the aesthetic imagination: gender blinks its way into the floating
space of android hermaphrodites, skin smooths out into Flash flesh
cut at the speed of syncopation, and the face itself floats away, not into
an aesthetics of facialization, but into something more indeterminate,
more tentative, more slippery in the codes. Plug into the data flesh of
Telebody and suddenly you are a twenty-first-century artificial life-
form streaming in the data nerve, fast exiting the blood bonds of
organic flesh, cut off from the comfortable and assuring shoreline of
the human sensorium. And you want to go back, and you need to go
ahead, and the sound is grafting itself inside your cyber-ears and the
images are tattooing the retina, and suddenly you might, if you're
lucky, or maybe not so fortunate, feel this skin rip at that point where
the genetic future undocks from the soon-to-be-dumped vessel of the
organic body. And if that happens then you might just get the queasy
feeling that *Telebody* has taken you to a space of retro-culture lag. Not
so much a lag between the speed of the techno-future and the slowness of consciousness as just the opposite. In *Telebody*, the artistic
vision of Steve Gibson drives the language of genetic determinism to
its point of aesthetic impossibility. An art of mutant sounds and
android images for slipstreaming the body into the recombinant
future.

  Telebody projects itself as the 'art of the human figure in the digital
world,' but it is actually exactly the opposite – an art of the digital
figure in the (last days of) the human world. Digital image capture, 3D
scanning, performers as 'metaphorical bio-geneticists': *Telebody* is an
art of light-time and light-space. Here, the electronic worlds of sound
and image achieve a new aesthetic relationship with animated image-
objects coming under the control of performed sound-objects. It is as if
the universe of electronic visuality with its images recombinant of
animated screens and distorted human figures released the ink black
filter of the electronic eye, falling under the spell of music deeply
informed by the rhythms of new biological discourse. Like sonic stem
cells in a radical experiment on the flesh/machine interface, the music
performance is the key: *literally*, of course, since the high-blast domain
of performed sound objects triggers perturbations of animated
screenal images; and *metaphorically*, because in perhaps the first
instance of 'regenerative music,' Gibson's sequenced 'halo' of splice/
edit/remix techno sounds is the key to understanding the floating

world of animated digital figures. In *Telebody*, sounds are injected directly into the sensorium of the electronic image – the global noosphere of the bio-gen body of the twenty-first-century future. Here, if the male and female figures appear first in human form, and, moreover, in classic perspectival poses, that is because *Telebody* witnesses the fast dissolve of the human figure into the skin of macromedia bodies, and the whirlpooling of ocular perspective into the delirious pathways and seductive reversals of an animated world triggered by the sonic environment of speed-performed sound-objects.

Telebody plays the *sound-images* of the recombinant mind. Neither aesthetically futurist nor nostalgic, this artistic performance is in the nature of an attentive cultural preconscious. Invoking once again the mythic function of art as both a probe of the (technologically) new and a data sensor of imaginative possibilities, *Telebody* anticipates the impact of biogenetics on the cultural nervous system. With a confident sense that the wetware imagination of the artist trumps the hardware of the scientist, *Telebody* compresses all the codes, present and future, of biogenetics – sequencing, splicing, editing, filtering, dynamic animation – into a controlling scripture of transgenic music that suddenly goes beyond the sphere of electronic aurality to precipitate a new universe of digital bodies recombinant. Here, in the role of a 'metaphorical biogeneticist,' the music performer is a code-breaker, test-driving the *cultural* genetic code to its point of delirious alteration. In *Telebody*, the skin of the organic body is stripped away; the electronic image speeds up into dynamic filter animation; the environments of sound and imagery are liquified by a new vision of the digital interface; and performance itself mutates into blue-blur motion.

Telebody is a *digital transformer* of bio-genetic flesh.

Tattoo Sound

It's a full moon, mid-October evening and I'm working my way down No Go Avenue to a gig at Club Gamma Ray. My electronic sensors are splayed wide open and my ears are on full alert. David Kristian, a technoblast composer, emailed to say that he was doing a gig with Zilon, sort of a recombinant tattoo showdown in which human flesh meets data sound. He warns me to wear earplugs because what I'll hear is going to hit like a laser pulse wave, or as David would say: 'If the vortex machine is working just right, it should ripple the flesh on people's faces and pancake the eyes like a hard wind blowin' across the skin water.' Cheap electronic thrills with a little body mutation thrown in for good data measure. I like that. In fact, I *need* that because my soul is riding high-octane empty, no disturbances, just this zombie land with the cynical joker in command.

Just down the street from Gamma Ray, a large, sort of disco-square-dance crowd spills out of Club Metropolis. David Bowie is in town for a show, and I guess I should be into Ziggy Stardust Recombinant time but I quick check my culture alert patches, see that the red lights are still flickering on the low-energy cycle, so I push right on through. Out of the corner of my vid monitor I see three guys up the street, bagged out in night-walking, triple-striped Adidas running suits, giving a couple of working women a real tough time. Seems they haven't got the money or the inclination to do the honourable thing and pay the bill for the body rent time, so they've decided to take their street-time pleasure in low-level, mouth squealing, mean-spirited harassment. Trapped hookers and a posse of snickering guys. Don't much like it, but I let that one file back to the memory banks, take a sharp left at the gargoyles and android cockroaches and road warrior swirling der-vishes blasted onto the Gamma Ray anti-architecture, and immedi-ately fall forward on my virtual face into the labyrinth of the future.

The show at Gamma Ray was advertised as an art-imitates-photoshop simscene. Zilon, a local performance artist who mostly 'performs' by tagging buildings with his trademark signature, was to be filmed live being tattooed by an electronic needle carving weirdly beautiful Egyptian hieroglyphs on his back and face and arms, the sounds of which were to be spliced, mixed, and recombined by the composer, David Kristian.

It all started innocently enough. A flesh/metal dark Gothic scene with Zilon, stripped to the waist, his back to the spectators, and a tattoo artist bearing down close with an electronic needle working his flesh, and all of this fed live to a hovering vid for large-scale screenal display. But what made the scene really interesting wasn't the body tattoo, but the *sound tattoo*. Over in the corner, Kristian wove and shuffled like some alchemical artist mixing up a medieval sound brew from the Vortex machine, mixing live samples of the sharp reverbs of the tattoo needle with files from deep in the android sound memory of hisses and pops and shrieks and statics and hammerhead noise blows to our listening skulls.

Sound surgery. Tattooed sound spliced and recombined with such intensity and speed that natural body rhythms just choked up and shut down, and your nervous system *became* the rhythm of the sound. Might like just to have sat there a bit alienated maybe, watching, mind-drifting, but it wasn't possible because the sound surgery forced your body to be an active mediation between the (flesh) tattooing and the sound (tattooing). I was part of the performance. Resynched, no longer really in control of my body, riding sound vector waves, low and high, of Kristian's music. Abuse music for abuse flesh.

Even Zilon spewed to the max-magnitude of the Vortex machine. His body was sound-pulsed by the remixed tattoo needle – the Zilon body trapped in a double Kafka tattoo machine: electronic needles for the flesh, electronic sound for the organs internal. Precision-aimed with laser intensity at different bodily organs – vortexing the kidneys, rewiring the bowels, fibrillating the heart, sound purging the spleen, splicing broken synapses in the nerve-net – the sound tattoo of David Kristian tingled the nipples, viagred the cock, jump-started the ears. Sometimes hardly audible, it grabbed your stomach with a deep-drone silent sound. Zero-sound, so low that the room fell silent except that the concrete began to shake. Or sound so high that it split your nerves, interrupted again and again with a jar-jar roar as the sound barrier of the auditory universe is broken. Kristian kills the barrier of sound with his Vortex machine.

A perfect evening for liquid sound that flesh-shifts the face as it needles the internal organs of the body wetware. Digital fingers across the chalkboard of the nerves pulse the body so that flesh dies, organs mutate, eyeballs bulge up and then flatten down supine smooth. You feel the nausea waves hit, and the sounds of chairs emptying as people panic and rush the exits. But somehow in the chaos my mind pops out feeling tranquil and serene in the blue sky of vortex music. I'm sitting there suddenly all by myself in an abandoned room with raw sounds of electronic needle-tattooing and blood dripping and big-screen black and red images of it all, and I know that I've come home to the techno-mutate future where I always longed to be and where my android body belongs.

I've been resynched into tattooed sound.

The Temptation of St. Anthony

I was in San Francisco, cyber-city on the Pacific, meditating on

McLuhan and the digital galaxy, thinking about McLuhan's privileging of experimental art as a way of scoping the techno-future. In
an out-of-the-way, no-name gallery deep in the looming shadow of
Silicon Valley, I stumbled on a small, but really haunting, exhibit of
experimental art that reminded me vividly of McLuhan's insight. The
exhibit was called *Techne*, and it consisted of works by San Francisco
techno-artists, artists who had all worked in the Valley as computer
engineers and multimedia designers, but who had then quit high-
intensity market tech for electronic art because they desperately
wanted to find a means of expression for the social dementia that they
all saw as one of the coming violences of a society driven by the will
to virtuality. This exhibit, *Techne*, was located in the midst of the
howling hype-machine of the California software mind, it was refused
any support, hardware or financial, by the industry because it actually
had something to say, and in my judgment it was a form of prophetic
art.

In this exhibit, there was one electronic art creation that haunted me,
then and now. It was called simply after the story by Gustav Flaubert,
The Temptation of St. Anthony, and it was by Elliott Anderson, an artist
who was a NASA flight simulation engineer. In *The Temptation of St.
Anthony*, you enter a darkened room, and see the image of a naked
man curled up in a fetal position on a white couch in the corner. This
image of the naked man exhibits all the ritualistic behaviour of the
obsessive-compulsive: aggressive movements, fractured vocal response, discomfort, a kind of taut-flesh anxiety just waiting to implode. An eerily beautiful hologramic simulation in which the naked
figure slowly turns, and almost mnemonically repeats the letters of the
alphabet.

And it's all a tech-panopticon. Not only the hologram of the obsessive-
compulsive, but the position of the observer is tracked by sensors.
Unseen and out of view, there is a controlling algorithm which stores
up in its memory bank a history of past observer positions, which
then trigger obsessive-compulsive behaviour. Thus, what you actually
see is the virtual flesh of psychosis, of real psychoanalytical decompo-
sition of the subject spliced and remixed by digital history. Yesterday's
audience for *The Temptation of St. Anthony* triggers today's behaviour.
My participation in this event-scene today will trigger the future of
the obsessive-compulsive. The overpowering feeling is sadness and
melancholy and a vivid sense of pain.

Max Weber once prophesied the appearance of the 'iron cage' as the
ruling image of uniform visual culture. Elliott Anderson prophesies

the appearance of the 'digital cage': a digital cage where obsessive-compulsive behaviour is the working norm for wired flesh: aggressive behaviour, fractured vocal responses in real-time conversations, discomfort at coming out of the wired cage. *The Temptation of St. Anthony* is a brilliant psychoanalytical diagnosis of the deep decompositions of the human subject in digital reality: a fluid, liquid electronic art that reflects deeply, that laments deeply, on what issues from the meeting of human flesh and digital reality.

The *Temptation of St. Anthony* is the dark outrider of *Understanding Media*, the artistic aporia that haunts the messianic enthusiasm of technotopia.

Infrared Sound

The digital body is a violent zero-point for the force-field of technology.

Tech drugs us from within, amuses us from without, serves us in robo-labour, speeds and sexes and teases us, pleases and despairs, facilitates and harms. It can rechannel eyes, rezone ears, and resynch nerves.

Tech high tides the mind. Enchanting and despairing, ennobling and humiliating, fascinating and boring, uplifting and deadening, tech violates and seduces.

Tech can make us wealthy or poor, give us life in the new economy or living death in dead-end work.

Digital tech as a labyrinth of seduction and domination, the dynamic momentum of which is preserved by the comforting belief in the possibility of actually stepping outside and transgressing the wired world. Wake up one day, ooze out of bed with your mottled flesh and a big hunger on for the solitude of the natural body. Yank electronic scuzzy ports out of the body's orifices, silence the TV, mute the radio, refuse CD and CD-ROMs, maybe even vinyl, and pass by all the computer games with a sigh of wistfulness. Be yourself again. Be natural. Out the media. Escape the digital zone. Refuse the game of technology.

Don Ritter's digital art destroys these last comforting illusions. In his electronic imagination, there is no possibility of stepping outside technology because tech *is* us: it's the air we exhaust, the body we drive, the spirit we advertise, the nerves we speed, the sound we recall, the eyes we object-orient. Tech is the self we possess.

Artist's Image-Capture

CAPTURED MOMENTS is a series of interactive video and sound installations which present visitors with four types of common interactions: interactions with people, interactions with nature, interactions with machines, and interactions with mass media. Sound and imagery occur continuously within these installations, even in the absence of visitors. When visitors are present, however, they instantly influence the installations. Specific sounds and imagery are presented according to the number, location, movement, lack of movement and temporal activity of the visitors. Within INTERSECTION, the sounds of cars moving across a dark space screech to a halt when a visitor walks into its path. And in TV GUIDES, a live television broadcast switches to the command 'Please Remain Still' when a visitor enters the space. When a visitor walks into the SKIES installation the imagery of a night sky projected onto a floor becomes a clear sky with a walking path.

When a visitor leaves one of these installations, sound and imagery return to the moment that existed prior to a visitor's presence. A visitor's impact is only temporary, having no lasting impact on the nature, machines or mass media contained within the installations.

Although a single visitor's presence and movement within the installations cause specific imagery and sound, the combined activities of all visitors will cause events which are not possible by a single visitor. The visitors, possibly strangers to each other, must co-operate to experience the works to their full potential.

The three installations have been structured according to the following relationships:

SKIES: interactions with nature: cooperation required

INTERSECTION: interactions with machines: fear results

TV GUIDES: interactions with media: actions controlled

Don Ritter (Artist's Statement)

The Invisible Blows of Technology

Captured Moments is intended to make us aware of the imminent dangers associated with techno-culture. In an age in which McLuhan's prophecy has come true, namely that under the stress of accelerated technological change, the sensorium of the body shuts down and goes 'numb' for survival, Ritter suggests a new counter-strategy – waking

the full array of the body's senses up in order to overcome a technological threat intent of reducing *us* to servomechanisms of a digital
future which we may have created but over which we have lost
effective control. The shock of waking up suddenly from amnesia is
painful, which explains, perhaps, the psychological dislocation of
perception provoked by Ritter's art. An artistic counter-strategy of
psychological dislocation for the body numbed by the invisible blows
of technology.

Heavy Traffic

Don Ritter's work Intersection stalls the body with sound. Cars racing
the freeway with no time to waste screech to a halt as visitors try to
cross eight lanes of virtual traffic.

You enter a darkened room in which all bodily senses except hearing
have been deliberately shut down. The sound heard is menacing yet
familiar – fast-moving freeway traffic, with the sounds of individual
cars audible by their rising scream as they approach and their low-
pitch roar as they pass the position of the observer. A violent digital
field, perfectly simulating the crash vectors of freeway traffic
life.Through a series of twinned speakers at each end of the room,
traffic sounds oscillate back and forth, wave-sound pulsing the high-
velocity body. Another accident of sound on the virtual freeway.

Don Ritter's art tells the story of what technology is doing to us every
day. His work makes visible the invisible force-field of technology. In
Heidegger's sense, Ritter 'presences' technology. We are always being
smashed by the freeway traffic of high technology. The overwhelming
feeling is one of fear of the techno-unknown: chance bodies, chance
sounds, chance accidents. Darkness hits, and you step off the curb.

Speed Flesh

With Don Ritter on my mind, I catch a ferry across the St Lawrence
River to body-merge with the high-speed, high-tech, high-energy
carnival crowd gathering at the Gilles Villeneuve Raceway for the
annual Formula 1 Grand Prix.

I know every inch of the fiber-optic-timed, crash curve raceway. During
the rest of the year, when the sun's up, the sky's dry, and the cops
aren't looking, it's where I put on my K2 HyperX-360 ultimate speed
in-line skates, de-armour my body with no better than bullet-proof

protective gear, strap on my no crush-skull helmet, and join a nomadic group of speed drifters who draft the raceway every waking minute of every skating day.

No ad bodies. Nobody speaks. Nobody knows your name, just your speed. Schumacher and Hill and Gilles's son, Jacques, have nothing over us. They might drive the straight-away at 300 kph and flash-curve the twisting raceway, but they're cocooned in a no-burn gas, no-hurt crash, total computer car, monitored and recorded and imaged and prodded and coached, in an anywhere, anyplace city, any country time and space, gravity-vectored, and folded away outside the natural elements.

Not us. When we nomad skaters take the curves, we're just a micro-wrong leg-muscle move from crashing, and when we skate the speedway in front of the empty grandstand, we don't need an optically registered timeboard to tell us that we've gone off the edge. Off the *speed* edge where our bodies begin to bend the curve of time, space, and fear.

Like those night-time skates in San Francisco where a group of us converge at the top of Telegraph Hill, do a quick equipment check of our skates, making sure, most of all, that the brakes have been safely removed, and then form a downhill skating cluster, V-shaped, bent down deep to catch the wind drafts, and then ramp towards the Bay far below, knowing all the while we're not going to die and we've never had it so good and we *are* speed and we hope, just hope, that no autos or bikes or cable cars or cats or dogs or alley rats or peds jam up the suicide run, taking us back to bodies skin-shaved on the hard concrete beach of night-time cool.

Flesh Tech Reboot

DNA Futures

It's 3:00 am universal on-line time and I've got the viral sounds of DJ Spooky wire-amping my good ear with my fade-sound ear searching the deep screech zones of the Web for a quick recombinant fix. An email tag-friend from the deep submerge scratch'n hide background of wireless Iran had broadcast this cryptic message before she was shut down by the purity police. Something about a new zone recombinant where the heavy-hyped future of the Human Genome

Future wasn't waiting around for government research funding and slow-time computer sequencing but has already mapped out the post-genetic future. Speed art for the overexposed body.

Looking for a quick twenty-first-century medicine-show cure, my fade-ear scratched out the code to this outlaw site. The colour was reverse transcriptase, sort of normal chromatics cut with the passion scent of the reptilian brain. Colours with tangible smells and hidden fears: digital wallpaper of floating body parts and genetic sequences. Sequenced artwork for recombinant eyes.

It's a go-go mandatory code site, and so you set the configuration of your virtual flesh to what you know must be the secret password, *Trinity 2*, and with just that passing sense that the laws of mythic repetition aren't going to change no matter what the body-type, the flash organic sea suddenly opens up and you're vectoring into a going-out-of-business party for body at the edge of tech flesh. Now, it might take a mention that everyone is crying for admission at the gates: all the diseased and the sick and the nervous and the healthy and the normal and the organ-peddlers and the immortality-seekers. Seems that the electric word had got out pretty fast about this new experiment in post-genetics, this 'bible to human life' to be found in a new cache of the dead flesh scrolls.

I hear a cheerful bah-bah and I just know that Dolly is there. And she is. Chewing some RNA grass in the corner of the virtual cell, obviously trying to look too smug about being the first animal flesh to cross over to the post-genetic side. But she's not alone. Because like a pilgrim's tale of the DNA reboot remix way, there's a lot of post-human flesh clogging the lanes. Guys with vector-flash heads and women with code for skin, sequences for eyes, and deep mutate for a tribal personality. And, of course, there all the trendy trans-mixers: part-flesh/part-scriptase – visually beautiful reconstructions of the old human cranium blown up to post-genetic performances. Francis Bacon'd body parts cut with Max Ernst in his Arizona desert days. But best of all are the crowd of post-biologics. Like a recombinant version of the parodic industrial architecture that quickly went out of fashion after the Beaubourg showed the real possibilities and sub-real limits of exposing the plumbing of buildings, the post-biologics dispensed with skin altogether, except sometimes as a translucent screen the better to highlight the disappearance of surface flesh. Here, the spinal cord tapped directly into spiralling nerve ganglia and floating organs, refurbished by weekly mud baths in the DNA scrubber. Post-biologic flesh happily on its way to an implosion of the limit-organs into pure spinal wetware.

Eye-Through Images

The Post-Alphabet Future

The real world of digital reality has always been post-alphabetic. Probably because the letters of the alphabet were too slow to keep up with the light-time and light-speed of electronics, the alphabet long ago shuddered at the speed of light, burned up, and crashed to earth. Writing can't keep up to the speed of electronic society. The result has been the end of the Gutenberg Galaxy and the beginning of the Image Millennium. Images moving at the speed of light. Images moving faster than the time it takes to record their passing. Iconic images. Special-Effect Images. Images of life past, present, and future as culture is fast-forwarded into the electronic nervous system. Images that circulate so quickly and shine with such intensity that they begin to alter the ratio of the human sensorium.

This is probably why artists, scientists, and engineers from Xerox Parc have created a new media installation titled *Experiments in the Future of Reading* (XFR) at the Tech Museum in San José, California. All the experiments in the future of reading project have a very practical purpose: to suggest new consumer products for post-alphabet society. Here, the alphabet is blasted apart and creatively reconfigured by the shock-wave of electronic culture. Touch-screens fill with texts which shift at any moment to follow another story line: single words that open up into continents of lost dreams; paragraphs that recombine into novellas; stories that compress into a single emotion. Or huge, gleaming light-tables on which are displayed graphic puzzles that can only be solved by physically tilting the table back and forth by hand, watching the letters of the alphabet slowly roll across the screen, forming new creative combinations. Literally, *hand-writing* for the new electronic cave-dwellers. A paradigm-shift in the form of ideas for new consumer products in which writing itself bubbles to the electronic surface, searches anxiously for its lost chain of (alphabetic) signifiers, dances hesitatingly across the old literary divide between metaphor and metonymy, finally realizes that words are on their own in a liquid digital world, and comes to life as light-through and sound-through and eye-through electronic words. The words slide up and down, mutate one to the other, creating new digital meanings. Pixel events, light-screen language, and soundscape texture.

Or, consider my personal favourite. A children's book telling the story of a cool cat doing the jazz scene in San Francisco. Except this time, rather than reading the book, *you play the reading*. Sit in a comfortable armchair equipped with micro-speakers (with a mega-computer tucked away behind the chair), open the book, run your fingers over the pages, and the sounds of jazz on the written page suddenly surround-sound your ears. The cool cat at the *Purple Onion*, at the *Hungry I*, or at an after-hours club down by the docks. In traditional reading culture, the eye was privatized, shut up inside the privacy of the central nervous system, isolated from the other senses. In the future of (electronic) reading, the eye goes public. It reconnects to the other senses, notably to the ear and the hand. *Tactile Reading*. Touch the page at any point and the sounds of jazz being written about can be instantly heard. You are actually in the sound field of the book. Move your hand closer to the page or further away, and the sound intensifies or fades accordingly. The end, therefore, of passive reading, and the beginning of in-depth participation in the electronic book. The future of reading will be fun. It will be experimental and immersive. It will be unpredictable. It is a full-body, full-mind, full-ear, full-eye experience. It will certainly involve the complete ratio of the senses. *Instantly, you are the reading*.

Or are you? If this project is about the 'future of reading,' then what's really being read? Not words rolling off light tables or books as soundscapes, but the *eye of human flesh itself*. Seduced by electronic reading as a packaged consumer product, the eye is externalized in the transcendent form of a light-object, a sound, a liquid consumer graphic, a simulacrum of ocular perception.

Virilio's 'sightless vision' or a game of alphabet soup?

Clicking-In to the Global Show

Did you catch *Quantum Project* on the Net? According to its promo, what the *Jazz Singer* did for the age of talking motion pictures, *Quantum Project* will do for the Internet as the global cinema.

Quantum Project is the Holy Grail of the tech future, that magical point where two previously separate media – cinema and the

Internet – touch and spark and converge. More than a made-for-TV movie in the *Matrix* mode, *Quantum Project* is the planet's first big-budget Hollywood-style made-for-the-Internet movie. Here, Hollywood crosses Silicon Valley, and the result is digital cinema with a big twist. Because what's really converging in *Quantum Project* is not simply two media – one millennium new, the other twentieth-century old – but something much more interesting. Here, the real software of Hollywood – its star system together with its high-intensity promotional culture – merges with the streaming software of the Internet to produce an Internet cinema that is global, immediate, and intense. When Hollywood promotional culture meets the planetary distribution system of the Internet, the result will be the world instantly retooled as a global cinema. When the world becomes a global show, the Internet will finally be experienced as popular consciousness. It will have its stars and its stories and its tragedies and its scandals and its blockbusters and its failures. The Internet will be the geist of electronic life. Going to the Internet will be the ticket to the future.

What Hollywood does best is streaming mythology with electronics, bundling charismatic stars and advanced (imaging) technology to produce a celluloid vision of life in the high-tech future. In these sometimes wonderful, sometimes haunting cinematic images, electronics is directly downloaded into the human imagination. For its sheer consumer appeal, nothing beats it. Cinema is iconic, fascinating, seductive, and, of course, often extremely profitable. Consumer electronics of a special sort blown up to the size of an IMAX screen. Maybe this is why the secret dream of all the Palm and PowerPCs and interface devices of the world of consumer electronics has always been to leave behind their purely instrumental work-day role as enablers of fast communication, becoming instead real players in the creation of human dreams – *interfaces to the stars*. Which is why *Quantum Project* can attract such a crackle of media excitement. Because what is really a quantum project is not just digital cinema, but the future of consumer electronics. Following the thread to the stars is the quantum project of the global show. Interfacing hot consumer electronics with cold cinematic stars is the future theatre of eyeball culture.

But, of course, digital cinema won't leave the Hollywood star system unscathed. Because let's face it: the real stars of digital reality are *special effects*. Cool software programs that realize impossible perspectives: special-effects sequences that can be so fascinating and seductive because they always deal with reality hyped-up to the point of hyper-reality. *Matrix* bodies moving faster than speeding bullets. *Star War* warp jumps. Morphed flesh. Streamed vision in every movie. Invisible digital editing in every televised newscast. And this is just the way it should be. In the age of the Internet, we are already living in a special-effects culture. Fast communication. Speed economy. Java memories. Linux open-architecture as a model for living by the dot.com generation.

The seduction of special effects is where the Internet has the jump on Hollywood. And this makes sense. Special effects is what digital cinema streamed on the Internet does best. The future stars of all the *Quantum Projects* of the future, therefore, as special-effects hybrids probably being dreamed up right now in the image-factories of the global cinema. Producing digital stars for the global show, therefore, as one future of electronic society. Not the *Jazz Singer,* but clicking-in to the Digital Eye.

9 Body and Codes

Codes of Replication

Consider the following remarks by Bill Joy, chief scientist of Sun Microsystems:

> The twenty-first century technologies – genetics, nano-technology and robotics – are so powerful because they spawn whole new classes of accidents and abuses.
>
> While replication in a computer or a computer network can be a nuisance, at worst it disables a machine or takes down a network service, uncontrolled self-replication in these newer technologies runs a much greater risk: a risk of substantial damage to the physical world.[1]

It's the same with Jean Baudrillard, who in *The Vital Illusion* prophesies:

> The specter that haunts genetic manipulation is the genetic ideal, a perfect model obtained through the elimination of all negative traits: no viruses, no germs, no defects ... The worst of it is that living beings engendered by their cwn genetic formulae will not survive this process of reduction. That which lives and survives by the codes will die by them.[2]

What is then the transgenic future? Are we on the way to the superhuman or to the subhuman? Recreating, redoubling, genetically twinning ourselves? What is the ethical future foretold by chimeras, hybrids and clones? by 'nuclear transplantation,' by new recombinant technologies of blastomere separation?

Baudrillard again:

> The question concerning cloning is the question of immortality. We all
> want immortality. It is our ultimate fantasy, a fantasy that is also at work
> in all our modern sciences and technologies – at work for example, in the
> deep freeze of cryonic suspension and in cloning in all its manfestations.[3]

What, then, is the fate of the human in the age of post-biologics?

Ironically, in a culture intentionally stripped of the language of myth, we
may be living in the most mythological of all times: the future of the *data
body, the transgenic body*, also mingles now with mythic stories forgotten
of minotaurs and monsters and aboriginal prophecies of the punishment exacted by the gods for transgressing the language of limits.

But remember:

That which is most analogically repressed also returns as the phantom absence which haunts the cold operations of the digital system.

That which is excluded will not be denied.

That is which rendered invisible always returns as the errant impulse
of the human heart, the forbidden imagination of the artist.

Without this return of reanimated memory and creative imagination,
wired culture will die of its own sterility.

Data is a natural cynic. Knowing no codes other than the software
ordinals of its own artificial universe, signifying nothing other than its
own telemetry, data streams in a weightless medium of zero-intensity.

Flowing invisibly, but no less violently, through the light-arrays of
the wireless world, data is the elemental material of the will to technology. Pooled in gigantic vats of cybernetic information, databasing is
what we mean now by history. Our electronically traceable past, our
digitally monitored present, empty data is the destiny of our artificial
future. THIS WIRED FUTURE DESPERATELY REQUIRES FOR ITS
SURVIVAL ARTISTIC DNA.

Codes of New Media Art as Reverse Engineering

To the question: Is it possible to reverse the momentum of new media
under the sign of wired culture? the answer is clearly affirmative.

Reverse engineering the new media has already begun. It's called
new media art. Its codes are threefold:

First, new media art reanimates a system which is dying of its lack of
creative energy with the repressed memory of that which has been

excluded, both from its analogue past and its electronic future.

Secondly, new media art fulfils McLuhan's vision of every margin a potential centre; every electronic periphery the galactic centre of a new media hub; every media externalization of the central nervous system an opportunity for radically altered human perception.

Finally, new media art is about *enhanced perception*. The aesthetic hype of new media art that parallels absolute technology is all about creating a totally immersive experience. I think this is a mistake. Mass media are effortlessly immersive. Numbing is what they do best. They fully colonize the human sensorium. They dominate perception. 'Sightless vision.' The point is not to mimic mass media aesthetics, but to break its spell. An art of enhanced perception, creating aesthetic conditions for a *fundamental attunement* to the world in which art struggles, an art of *attenuated awareness*. The code of enhanced perception is not new in the history of art. It's Francis Bacon's smeared body triptychs; Valasquez's painterly parodies of the theatre of representation; Scanner's sound scans of the culture of cities; the artistic rebellions of dadaism, surrealism, fluxus, automatic writing for a chaotic time. Not an absolute interactive art for absolute technology, but art at the meridian. New media art which opens up awareness of paradoxes of sound, narrative, images, bodies on the one hand while avoiding being crushed by the numbing weight of mass media aesthetics on the other. Enhanced artistic perception for a disabled digital space.

Today, new media art is a creative hub in which repressed images of hybrids, clones, chimeras flow back into a digital system that, for all its technicity, hardware, and software, remains fascinated most of all by that which it has disappeared: *the wetware of cultural imagination.*

Until now, the bias of new media has clearly, and perhaps necessarily, been towards creative engineering developments in hardware and software. That's the basic code of the digital bible: speed, transparency, connectivity, immediacy. Not for much longer though. *Like all media before it, the bias of digital communication is about to flip.* Paradox is about to trump transparency. Connectivity is about to be flipped by enhanced digital perception. Speed is about to be reversed by a new art of fast images and slow perception. Reverse engineering body and codes will be carried out by new media artists: the real avatars of the digital, and perhaps genetic, future.

Reverse Engineering

In computer software language, reverse engineering refers to the decompiling and disassembly of redistributable codes. Hacking system operating software to recover lost source codes, to migrate applications to a new hardware platform, to translate code written in now obsolete languages, debugging fragmented systems. So far so good: software engineering with a clear purpose: dreams of data flow analysis, control flow analysis, perfect system transparency, perfect software accountability, computer programming in the service of perfect telemetry. The Borg is smiling.

But in actual hardcore computer hacking, the original dream is flipped. Reverse engineering is turned into its opposite. Legions of critical hackers interested in aesthetic and political issues such as electronic freedom, enhanced interactivity for everyone, the global village as a shareware future, a Linux for every electronic brain, begin to reverse engineer the source codes of infoculture itself. They organize themselves into new electronic tribes. You can find them in hidden data valleys in the Net: the Reality Hacking Lab, Anonymous Lab, or perhaps the Software Redemption Lab, the Debugging Lab. They have their own favourite musicians: Scanner, DJ Spooky, Steve Gibson. They have their own new media heroes: the reengineered body of Stelarc, the reengineered visual imagination of Perry Hoberman. Their code work transfoms the terms decompiling, disassembly, and debugging into new media art practices.

And why not?

The speed of global technological change is transforming contemporary society. In a global culture driven forward by dramatic developments in technology, no aspect of politics, culture, and society is left undisturbed. The 'new economy' associated with computer technology simultaneously challenges the stability of the so-called 'old economy' of industrial capitalism and is then itself undermined by what Schumpeter once the 'creative destruction' characteristic of market-driven economies. Creative developments in digital education and improvements in 'distributing' digital knowledge promise to transform Marshall McLuhan's vision of 'education without walls' into a long-distance, but real-time, learning reality. Fundamental changes in the nature of communication, from Palms to a whole array of wireless devices, alter the media landscape of how and why and where we communicate with one another. More than technologies of production and communication, the digital future also delivers powerful technologies of consumption in the form of homogenous consumer culture, from 'branded' consumption to the global sound of hip-hop to instant membership in the digital generation. Today, in ways more pervasive than we may suspect, MP3, Google, Linux, and Zero-Knowledge are the real world of digital culture. What was once only a science fiction dream, wired culture has now become the operating system of increasingly mediated societies. Are we data flesh? Are we Telebodies?

Nowhere is this more evident than in the emergent technologies associated with the 'biotech future.' Biochips, cloning, stem cell research, transplantation, genetic engineering and mapping, transgenic humans, plants, and animals, reviving endangered species in animals bred expressly for that purpose: these are some of the trends in biotechnology that are profoundly influencing current attempts to 'design' the future of evolution itself. The future as the Genetic Matrix.

A major aspect of a political ethics relevant to the 'question of biotechnology' has to do with accurately forecasting the direction of change in the biotech future. What is particularly striking in the project of biotechnology is the very future of the body. In sharp contrast to contemporary understandings of the concrete, integrated body, the thrust of biotechnology is directed at a radical transformation of perceptions of the body: sometimes from the outside (cloning, xenotransplantations) and sometimes from the interior of the body itself (the Human Genome Project). This radical transformation of the body of the future is nowhere more explicit than in contemporary initiatives in biotechnology directed at the creation of the 'transgenic body' – literally an 'interspecies' body produced by genetically streaming animals, plants, and humans. Here, beginning with animal experiments undertaken by private companies, including Advanced Cell Technology, Infigen, and Geron Inc., the future body is envisioned as a product of genetic sequencing and inter-species transplantation. While the ethical implications of the 'transgenic body' need to be fully explored in the social sciences and humanities, new media artists have developed significant artistic representations of the biotech body of the future. In their artistic imagination, the image of the transgenic body is not only visualized, but is actually 'performed' in the new media codes of sound and images. Reverse engineers of the transgenic body, new media artists sometimes detect fatal flaws in its coding, perceptual and ethical, decades in advance.

There was a newspaper report recently of a meeting of the American Astronomical Society where astronomers reported the discovery of *huge voids*, or *'ghost cavities'* foaming through the surrounding gas of far distant galaxies – *'the aftershocks of ancient explosions ignited perhaps by matter falling into a black hole.'* These ghost cavities can't be seen with normal human vision, *but only with X-ray vision* – the ability to register extremely high-pitch electromagnetic waves, and thus detect ultraviolet frequencies.

Now I think that new media artists are like that: *they have X-ray vision,* a way of comprehending frequencies of the human condition where painful repressed memories are allowed to animate contemporary culture. Theirs is an art of 'ghost cavities,' an art of huge voids in the political galaxy that, once released, flash their way onto human perception like aesthetic hooks in human memory.

Body and Codes

In his book *Technics and Civilization*, Lewis Mumford said that fundamental shifts in the direction of technology are often preceded, sometimes by centuries, by prior transformations in culture, that in history, as in life, culture anticipates technology. For example, in Mumford's view, the industrial revolution did not really begin with the modern age, but was actually set in place by fundamental cultural transformations in the medieval age. Not just the mechanization of time and changes in property relations, but, more decisively, the model of the factory was anticipated centuries in advance by the model of the monastery, with its routinization of time, careful regulation of daily activity, disciplining of labour time, and specialization of function.

Now while Lewis Mumford wrote a brilliant cultural history of the technological past, what of the technological future? We are transitioning between the end of the modern age and the beginning of a technoculture which has two dynamic spearheads: utopian dreams of a wireless future and radical experiments in biotechnology. In the 1960s, McLuhan could only think projectively about a world organized like a 'global brain,' but for us data bodies, networked intelligence, streamed music, wireless communication, business at the speed of the database are the culture of the digital everyday. In the same way, dramatic biogenetic experiments in cloning, somatic cell therapy, xenotransplantation, organ harvesting, this whole redeployment of Darwin's *The Origin of Species* in the new language of the Human Genome Project will migrate rapidly from the periphery of human attention to its centre. The creation of *Transgenics* – beings part-human/part-animal/maybe part plantlike – eyes that allow us to see the sky in terms of patterns of gravitational waves like migrating birds, genetically modified hearing at the level of previously invisible sounds, is the likely new order of values of the twenty-first century. The ethical consequences of this cultural change are decisive since what is challenged today is not simply the goals of technology but what it means to be human, to be post-human, to literally have the mind interfaced to the speed of digital networks, to live at the lip of the Net, to have the body subject to a future of genetic modification.

In terms of a strict technological determinism, we are probably already living in a cloner culture in which dreams of xenotransplantion (cloning animals for organ harvesting), biopharmacology (those vast pharmaceutical factories of artificially bred animals for the manufacturing of new drugs), and creating transgenics are the dynamic momentum pushing technology at the speed of bio-business forward. *But are we ethically prepared for this?* Are we suiciding ourselves to virtual life? The law of eternal recurrence will not be denied. We do not have a special exemption. The harvesting of other species will inevitably have its reverse side: the harvesting of the human species at the lip of the Net. Maybe the larger cultural discourse that we are presently caught up in is that digitality and biotech are the chosen mechanisms by which human beings will be interfaced to the data world, genetically modified for more perfect system transparency. A culture of disappearances.

Have we really thought through the relationship between *first wave eugenics* (Darwinian naturalism, Spencer's social theory of the survival of the fittest), *second wave eugenics* (those theories of racial improvement through selective breeding which were practised in most western countries, including Canada, the United States, and most traumatically, Nazi Germany), and *third wave eugenics*, which is popularized today under the sign of the Human Genome Project, and other big science experiments in the field of biotechnology and bio-pharmacology with their delirious schemes for the post-human body in a post-evolutionary time. Are we really ready for the culture of the double? Do we really understand the fate of the transgenic body?

The actual situation today may be that we are living in a culture of twenty-third-century engineering and nineteenth-century ethics. We experience a new media cultural lag in which the speed and intensity of technological change seemingly overwhelm human perception and ethical reflection. But if Lewis Mumford is correct that culture always anticipates and actually outstrips in imagination the hardwired realities of the technological future, then it's probably the case that the biotech future is already being 'worked through' in the human mind by those brilliant outriders of the future: the artistic imagination. Their imagination reverse engineers the future. It is the basis of a new ethics equal to the incommensurable changes precipitated by biotechnology.

For example, consider contemporary film, where the consequences of the present trends in technoculture are worked through relentlessly and creatively.

Gattica explores the reality of the new eugenic class, a world sharply divided into the eugenically privileged and the new dispossessed. In an age of anxiety over the fate of the body, it also probes the tragic dimensions of a new circumstance of life in which we are constantly betrayed by our bodies: betrayed by physical weaknesses that disqualify us, betrayed by bodily fluids that are used by the state in *Gattica* for constant surveillance of the population. Nanosurveillance.

AI, this strange transgenic cinematic hybrid of Kubrik and Spielberg, with its necessary doubled ending – one cold-eyed dystopian and the other upbeat sunny California utopian – probes the cloner future by asking that fundamental human question: Can the pure, unequivocal love of an artificial life form – a clone – for his mother be reciprocated? Or is there some ancient prejudice as to the relationship between the natural and the artificial which will always prevent true reciprocity – the basic starting-point of any authentic human relationship?

Or consider *Memento*, which makes amnesia a working condition of life in hyper-reality and reduces experience to an empty image-repertoire of snapshots. Whom do you trust? Whom do you betray? What's life really like in the simulacrum cut with Hobbes's 'war of all against all'?

Or what about already classic films such as *The Matrix* with its gnostic appeal to the new body in the technological sublime: Will that be a red pill or blue pill? a life of amnesic slumber – eyes wide shut – or waiting for the Messiah who dodges the fatal bullet at the speed of special-effects photography?

But isn't there always a third choice? Like *Fight Club*, this cinematic version of the suicide clubs that began in the 1970s in Seattle and Portland, where life suicides itself into doubled selves and hunting pack relationships and the ecstasy of 'pure hatred'? A film about the culture of nihilism: the cultural fallout of advanced technological society. Or think of those other brilliant independent films, *Green* and *Judy Berlin*. They are both prophetic visions. *Green* takes place in Phoenix,

Arizona, under the hot, almost suffocating, noonday sun, a Nietzschean film of four ex-college students driftworking together across another day of their life, with this feeling of the absurdity of things in the air with a lot of quiet despair. Or *Judy Berlin*, that's suburbia as a dead terminal of high technology. A film student living at home in Babylon, Long Island meets up with a high school actress wannabe who is about to decamp for Hollywood. All this takes place in the twilight of an eclipse of the sun. You can just feel in *Judy Berlin* the subjectivity of the culture of boredom. If technology means, as Michel Foucault claimed, technologies of consumption, of communication, of subjectivity, of culture, then contemporary film is a laboratory of the future, predicting decades in advance the paradoxical consequences of new technologies and, more decisively, providing a visual and emotional language by which to think our present destiny as we seemingly fall upwards into the skyless space of the electronic future.

This is even more the situation with new media art. Necessarily, its cultural creativity parallels the creativity of the biotech future itself. Paradoxically, as a new artistic medium it must of necessity create forms of aesthetic expression that are simultaneously deeply interior to the language of technology while remaining somehow exterior to and independent of the destiny of technology. In philosophical terms, new media art is simultaneously a form of art that is in-itself and for-itself, *en soi* and *pour soi*, both a reflection on and within electronic culture, and yet a form of aesthetic expression that gains cultural autonomy by transforming the language of electronic culture from its present obsession with the practicalities of techne to the deeper dreams of poeisis. New media art is the poetics of electronic culture. And as the poetic imaginaire of electronic culture, new media art is both a way of projecting into the future and a very real circling back to heal an ancient separation that began the story of modern technology, the radical separation of technology into tools untouched by the paradoxes of human imagination and art divorced from the power of scientific practice. *The engineer and the artist: an old story that has been told again and again. C.P. Snow's Two Cultures, Northrop Frye's 'educated imagination,' Jean-François Lyotard's vision of art as a 'transformation matrix,' Martin Heidegger's sense of art as a poetics of the technological imagination.* But if, as Albert Camus says in *The Myth of Sisyphus*, myths aren't made to be memorized but to be *reanimated* by having our own visions, our own burning questions, blown into them, then what is the aesthetic situation today in this new age of bio-engineering the future and art in the age of new electronic perception? What is the place of the engineer and the artist in the contemporary culture of Body and Codes?

Until now digitality has operated under the sign of globalization. The multimedia future has been centralized, hardware oriented, unidirectional, moving at the speed of light, relentlessly striving for greater connectivity, immediacy, accessibility, haunted by the phantom of the last mile, larger bandwidth, panicked by network congestion, offshore viruses, the spectre of dark fibre, the always missing convergence of computers, music, television, and film. The wired *Canarie* that doesn't sing. Convergence without a sustaining meaning. Like browsers with nowhere to go on a digital stroll.

But as in life so too in digitality. Every presence has its absence. Every sign has its reversal. Every new medium has its hidden suppressed side. New media quickly establish themselves in the language of facilitation: the wired utopia of connectivity, speed, big bandwidth. But every new medium only really consolidates itself, only really comes alive, in the language of its phantom absence. Without art, digitality perishes. Art is the essential survival strategy of digitality today, and perhaps the basic survival strategy of human life itself.

Notes

1. The Will to Technology

1 Hannah Arendt, *The Life of the Mind* (New York: Harcourt, Brace, Jovanovich, 1978), 50.
2 Don DeLillo, 'In the Ruins of the Future,' *Harper's*, December 2001, 33–40.

4. Hyper-Heidegger: The Question of the Post-Human

1 Martin Heidegger, 'The Word of Nietzsche,' in *The Question Concerning Technology and Other Essays*, translated and with an introduction by William Lovitt (New York: Harper and Row, 1977), 109.
2 Martin Heidegger, *The Fundamental Concepts of Metaphysics: World, Finitude, Solitude*, translated by William McNell and Nicholas Walker (Bloomington and Indianapolis: Indiana University Press, 1995); *The Question Concerning Technology and Other Essays; An Introduction to Metaphysics* (New Haven and London: Yale University Press, 1987); *Discourse on Thinking*, translated by John M. Anderson and E. Hans Freund (New York: Harper Torchbooks, 1969); *Being and Time*, translated by John Macquarrie and Edward Robinson (San Francisco: Harper and Row, 1962); *The End of Philosophy*, translated by Joan Stambaugh (Chicago: University of Chicago Press, 1973).
3 Martin Heidegger, *Nietzsche, Volumes III and IV*, 'The Will to Power as Knowledge and as Metaphysics,' and 'Nihilism' (San Francisco: Harper, 1987).
4 'Enframing means the gathering together of that setting-upon which sets upon man, i.e., challenges him forth, to reveal the real, in the mode of ordering, as standing-reserve. Enframing means that way of revealing

which holds sway in the essence of modern technology and which is itself nothing technological' (Heidegger, *The Question Concerning Technology and Other Essays*, 20).

5 Heidegger, *Discourse on Thinking*, 45.

6 Ibid., 54–5. 'But if we explicitly and continuously heed the fact that such hidden meaning touches us everywhere in the world of technology, we stand at once within the realm of that which hides from us, and hides itself just in approaching us. That which shows itself and at the same time withdraws is the essential trait of what we call the mystery. I call the comportment which enables us to keep open to the meaning hidden in technology, openness to the mystery.'

7 Ibid., 55. 'What could be the ground for the new autochthony? Perhaps the answer we are looking for lies at hand; so near that that we all too easily overlook it. For the way to what is near is always the longest and thus the hardest for humans. This way is the way of meditative thinking.'

8 Heidegger, *The Question Concerning Technology and Other Essays*, 4. 'Likewise, the essence of technology is by no means anything technological. Thus we shall never experience our relationship to the essence of technology so long as we merely conceive and push forward the technological, put up with it, or evade it.'

9 Ibid., 'The Age of the World Picture,' 115–54.

10 Heidegger, *Discourse on Thinking*, 52.

11 Ibid.

12 In this text, Heidegger provides the theory of completed nihilism: its fundamental attunement – 'profound boredom'; its method – the disciplinary practices of biogenetics; its dominant cultural sign – terminal drifting towards generalized 'indifference.'

13 See, in particular, Heidegger's reflections on the historical destiny of the German 'folk,' in his *Die Selbstbehauptung der deutschen Universitat*, 'Rektoratsrede' (Breslau: W.G. Korn, 1933).

14 Heidegger, *The Question Concerning Technology*, 'The Word of Nietzsche,' 102. 'In the willing of this will, however, there comes upon man the condition that he concomitantly will the conditions, the requirements, of such a willing. That means: to posit values and to ascribe worth to everything in keeping with values. In such a manner does value determine all that is in its Being.'

15 Heidegger, *Nietzsche*, 'The Will to Power,' 197. Beyond the question of technology, Heidegger argues that the will to will that is the essence of technological destining always requires that human and non-human nature be reduced to the function of 'standing-reserve.' Thus, for example,

in Nietzsche, Heidegger describes the essential movement of the will to power as gathering into itself means for the 'preservation' of power. 'Therefore, enhancement of power is at the same time in itself the preservation of power.' It is in this sense that Heidegger describes the technical condition of human subjectivity as 'standing-reserve' in *The Question Concerning Technology and Other Essays*, 23. In his essay 'On the Question of Being,' Heidegger notes: 'The reduction that can be ascertained within beings rests on the production of being, namely, on the unfolding of the will to power into the unconditional will to will.' Martin Heidegger, *Pathmarks*, edited by William McNeill (Cambridge: Cambridge University of Press, 1998).

16 Martin Heidegger, *Basic Writings*, edited by David Faarrell, Krell (San Francisco: Harper, 1993), 'The Origin of the Work of Art,' 140–212. For Heidegger, the importance of art in the technological milieu was precisely to open the question of technology to a different form of interpretation, not only the logic of 'calculability' but also the revelation of poetry.

17 Heidegger, *Pathmarks*, 258. 'Homelessness so understood consists in the abandonment of beings by being. Homelessness is the symptom of the oblivion of being. Because of it the truth of being remains unthought.'

18 Ibid; 'What Is Metaphysics,' 93. 'Being held out into the nothing – as Dasein is – on the ground of concealed anxiety makes the human being a lieutenant of the nothing.'

19 Heidegger, *The Question Concerning Technology and Other Essays*, 44.

20 Ibid., 100. In 'The Word of Nietzsche,' Heidegger draws the conclusion from technological objectification as destiny: 'Man, within the subjectness belonging to whatever is, rises up into the subjectivity of his essence. Man enters into insurrection. The world changes into object. In this revolutionary objectifying of everything that is, the earth, that which first of all must be put at the disposal of representing and setting forth, moves into the midst of human positing and analyzing. The earth can show itself only as an object of assault, an assault that, in human willing, establishes itself as unconditional objectification.'

21 Ibid., 48.

22 Heidegger, 'The Turning,' in *The Question Concerning Technology and Other Essays*, 41.

23 Heidegger, *The Fundamental Concepts of Metaphysics*, 162. 'Profound boredom, its being left empty, means being delivered over to beings' telling refusal of themselves as a whole. It is thus emptiness as a whole.' Intensifying Nietzsche's admonition that man has grown tired of himself, Heidegger asks: 'Has man in the end become boring to himself? – as the question

in which we ready ourselves for a fundamental attunement of our Dasein'
(161).

24 Writing of the 'grounding-attunement,' Heidegger states: 'In the first
beginning: deep wonder. In another beginning: deep foreboding.'
Heidegger, *Contributions to Philosophy (From Enowning)*, translated by
Parvis Emad and Kenneth Maly (Bloomington and Indianapolis: Indiana
University Press, 1999), 15.

25 Heidegger, *The Question Concerning Technology*, 20.

26 Ibid., 17.

27 Ibid., 16.

28 Ibid. 34–5. For Heidegger, the alternative to technology as calculative
reasoning lies in a vision of technology that has withdrawn into forgetfulness, namely that once 'the poeisis of the fine arts was also called techne.'

29 Ibid., 20–1. 'Enframing means the gathering together of the setting-upon
which sets upon man, i.e. challenges him forth, to reveal the real, in the
mode of ordering, as standing-reserve. Enframing means that way of
revealing which holds sway in the essence of modern technology and
which is itself nothing technological.'

30 Ibid., 33.

31 Ibid.

32 Ibid., 34.

33 Heidegger, *The End of Philosophy*, 100.

34 Heidegger, *Nietzsche*, 8. 'In the thought of the will to power, Nietzsche
anticipates the metaphysical ground of the consummation of the modern
age. In the thought of the will to power, metaphysical thinking itself
completes itself in advance. Nietzsche, the thinker of the thought of the
will to power, is the *last metaphysician* of the west.' Or, as Heidegger argues
concerning his general theorization of *completed metaphysics*: 'Technology is
completed metaphysics. It contains the recollection of techne and, at the
same time, the name makes it possible for the planetary factor of the
completion of metaphysics and its dominance to be thought without
reference to historiographically demonstrable changes' (*The End of
Philosophy*, 93).

35 Ibid.

36 Ibid., 92.

37 Ibid.

38 Ibid., 85.

39 Ibid., 100.

40 Ibid., 101.

41 Ibid., 102.

42 Ibid.
43 As Heidegger states: 'The end of philosophy proves to be the triumph of the manipulable arrangement of a scientific-technological world and of the social order proper to this world. The end of philosophy means the beginning of the world civilization that is based upon Western European thinking' (*Basic Writings*, 'The End of Philosophy and the Task of Thinking,' 435). The end of philosophy, then, as the beginning of a metaphysics of being that hovers around its own emptiness and on behalf of which it accelerates into the 'incessant frenzy of rationalization' and the 'intoxicating quality of cybernetics' (ibid., 449).
44 Heidegger, *The End of Philosophy*, 102.
45 Heidegger, *The Fundamental Concepts of Metaphysics*, 7.
46 Ibid., 163.
47 Ibid., 163–4.
48 Heidegger, *The Fundamental Concepts of Metaphysics*, 7.
49 Heidegger, 'What Calls for Thinking,' *Basic Writings*, 381–2.
50 Ibid., 382.
51 Heidegger, 'The Origin of the Work of Art,' *Basic Writings*, 197.
52 Carl Nolte, *San Francisco Chronicle*, 10 March 2000.

5. In a Future That Is Nietzsche

1 Friedrich Nietzsche, *Thus Spake Zarathustra*, translated by R.J. Hollingdale (New York: Penguin, 1978), 161.
2 Friedrich Nietzsche, *On the Genealogy of Morals*, translated by Walter Kaufmann (New York: Vintage Books, 1989), 61.
3 Ibid., 57.
4 Ibid.
5 Ibid.
6 Ibid., 58.
7 Ibid.
8 Ibid., 59.
9 Ibid., 66.
10 Ibid., 67.
11 Ibid., 65.
12 Ibid., 113.
13 Ibid., 84.
14 Ibid.
15 Ibid., 85.
16 Ibid., 87.

17 Digital ressentiment is the hypermodern sensibility of the 'last will': the restless will that would always prefer to 'will nothingness rather than not will at all.' Nietzsche's description of the last will is the essence of completed nihilism, that moment when the will turns into a haze of *surplus positivity.* The third essay of the *Genealogy* thus completes the thesis of Heidegger as the harbinger of completed nihilism, the thinker who extracts from Nietzsche's concept of the last will the will to 'profound boredom.'

18 Nietzsche, *Genealogy*, 43.

19 Nietzsche, *Thus Spake Zarathustra*, 0.

20 Nietzsche, *Genealogy*, 86.

21 Ibid.

22 Ibid.

23 Nietzsche, *Thus Spake Zarathustra*, 153.

24 Ibid., 43.

25 Ibid., 191.

26 Nietzsche, *The Will to Power*, 346.

27 *New York Times*, 23 June 2000.

28 Nietzsche, *Genealogy*, 86.

29 William Gates, *Business @ the Speed of Thought: Using a Digital Nervous System* (New York: Warner Books, 1999), 23.

30 Ibid., 37.

31 Ibid., 6.

32 Ibid., 7.

33 Ibid., 15.

34 Ibid., 449–50.

35 Ibid.

36 Friedrich Nietzsche, *The Will to Power*, 346.

6. Streamed Capitalism: Marx on the New Capitalist Axiomatic

1 Karl Marx, *Capital: A Critique of Political Economy*, translated by Ben Fowkes (London: Penguin Books, 1976); *Grundrisse: Foundations of the Critique of Political Economy*, translated with a foreword by Martin Nicolaus (London: Penguin Books, 1973).

2 Marx, *Capital*, 988.

3 Ibid. Not simply the 'doubled movement' of use-value and exchange-value as antinomic moments in the 'circuit of circulation' of capitalist production, but also 'objectified' and 'living labor' as the mirror of the flesh necessary for the valorization of the process of capitalist value itself. See

particularly Marx's account of the circuit of (capitalist) circulation theorized in the form of the 'Results of the Immediate Process of Production,' *Capital*, 948–1084.

4 For an application, see Marx's description of exchange value through 'the process of circulation' (*Grundrisse*, 235).

5 Refusing the 'value-form' at the centre of the circuit of capitalist production the status of a referential finality, the language of capitalism as a 'value-form' is the precise way in which virtuality displaces materialism as the code of capitalist production.

6 While Marx argues that the 'fetishism of the world of commodities arises from the peculiar social character of production which produces them,' he might also have noted that the fetishism of the world of production also results in the peculiar social character of the process of commodification, namely that commodities themselves are the legacy code for the 'value-form' of capital and, consequently, the first (material) objects to disappear into virtualities (*Capital*, 'The Fetishism of the Commodity and Its Secret,' 163–77).

7 Karl Marx, *Grundrisse*, 'The Chapter on Money,' 113–238.

8 Heidegger's theorization of the essence of technology in the language of 'enframing' and 'appropriation' perfectly describes Marx's analysis of the essence of capitalism as the alienation of surplus-value. In this case, Marx's concept of surplus-value is equivalent to Heidegger's diagnosis of the 'standing-reserve' as the alienated product of the process of technological production. What Heidegger *thinks* metaphysically in terms of the origins and consequences of calculability, Marx theorizes economically. In each case, we are presented with the ascendant 'value-form' of virtuality.

9 'The need of a constantly expanding market for its products chases the bourgeoisie over the whole surface of the globe. It must nestle everywhere, settle everywhere, establish connexions everywhere. The bourgeoisie has through its exploitation of the world-market given a cosmopolitan character to production and consumption in every country.' Karl Marx, *The Communist Manifesto* (New York: W.W. Norton, 1988), 58.

10 For a full description of the virtual class in relation to the digital commodity-form, see Arthur Kroker and Michael A. Weinstein, *Data Trash: The Theory of the Virtual Class* (New York: St Martin's Press, 1994).

11 Marx, *Grundrisse*, 253.

12 The specific importance of the 'Results of the Immediate Process of Production' is its analysis of the speed of the 'circulating commodity' as it exits the process of production and enters its definitive phase as the process of 'value valorizing itself' (Marx, *Capital*, 948–1084).

13 For a further elaboration of the ' hinge experience,' see Jean-François Lyotard, *Duchamp's Trans/Formers*, translated by I. McLeod (Venice, CA: Lapis Press, 1990).

14 Marx, *Capital*, 965.

15 Marx, *Grundrisse*, 253.

16 Marx, *Capital*, 953.

17 Marx's theorization of the process of capitalist realization begins with the thesis that 'commodities are the first result of the immediate process of capitalist production' (*Capital*, 974).

18 Ibid., 954.

19 Ibid., 955.

20 The commodity can be a 'universal, elementary value-form' because it represents the 'immanent unity of use-value and exchange-value,' that is, the 'immediate unity of labor and the valorization process' (Marx, *Capital*, 978).

21 Ibid., 990.

22 Ibid., 1008. For a discussion of Marx's theory of the disciplinary production of labour, see the section in *Capital* that deals with the reduction of labour to a 'quantum of value,' that is, 'a specific mass of objectified labour, to suck in living labour in order to increase and sustain itself ... Capital utilizes the *worker*, the *worker* does not utilize *capital*, and only articles *which utilize the worker* and hence possess *independence*, a consciousness of the will of their own in the capitalist, are *capital*.'

23 Jean-François Lyotard. *Economie Libidinale* (Paris: Minuit, 1974).

24 Jean Baudrillard, *The Mirror of Production*, translated by Mark Poster (St Louis: Telos Press, 1975).

25 Jean Baudrillard, *Symbolic Exchange and Death*, London: Sage, 1993, 2.

26 For a critical theorization of the significance of the 'rational terrorism of the code' as the third term that substitutes itself for the alternation of use-value and exchange-value in the immediate process of capitalist production, see Jean Baudrillard's essay on 'Marxism and the System of Political Economy' in *The Mirror of Production*, 111–67.

27 'The *functions* fulfilled by the capitalist are no more than the functions of capital – viz. the valorization of value by absorbing living labour – executed *consciously* and *willingly*. The capitalist functions only as *personified* capital, capital as a person, just as the worker is no more than *labour personified*' (Marx, Capital, 989).

28 Ibid., 985.

29 Ibid., 989.

30 Ibid.

7. The Image Matrix

1 Sarah Boxer, *New York Times*, 15 April 2001.

8. The Digital Eye

1 Octavio Paz, 'The Prisoner,' *Early Poems: 1935–1955* (Bloomington: Indiana University Press, 1973), 89–93.
2 Roland Barthes, *The Pleasure of the Text* (New York: Farrar, Straus and Giroux, 1975), 49.
3 Paul Virilio, *Open Sky*, translated by Julie Rose (London: Verso Books, 1997).
4 Sue Golding, *The Eight Technologies of Otherness* (London and New York: Routledge, 1997).
5 Ibid., xii.
6 Jim Robbins, *New York Times*, 2 March 1999.
7 James Weiser and John Seely Brown, 'The Coming Age of Calm Technology,' Xerox Parc, 5 October 1996, http://www.ubiq.com/hypertext/weiser/acmfuture2endnote.htm. The paper is a revised version of Weiser and Brown, 'Designing Calm Technology,' PowerGrid Journal V 1.01 (http://powergrid.electriciti.com/1.01 [July 1996]).
8 Ibid.
9 Ibid.
10 Ibid. This concept is elaborated in the section of the article titled, 'Calm Technology.'
11 Ibid., '(M)oving back and forth between the two' is theorized in the section of the paper titled 'The Periphery.'
12 Ibid. The technological possibility of transforming the ocular practice of 'centering and periphery' into a brilliant design concept is further elaborated in the section titled 'Three Signs of Calm Technology.'
13 Ibid.
14 Jean-François Lyotard, *Duchamp's Trans/Formers*, translated by I. McLeod (Venice, CA: Lapis Press, 1990).
15 Ibid., 27–8.
16 For background information on British Telecom's *Soul Catcher* project, see http://www.mindcontrolforums.com/mindnet/mn187.htm.
17 Michael D. West, CEO, Advanced Cell Technology, as reported in the *New York Times*. For a fuller account of the technological project of Advanced Cell Technology, see http://www.advancedcell.com/.
18 'Molecular Breeding,' *New York Times*, April 1999.

19 Ibid.
20 Ibid.

9. Body and Codes

1 Bill Joy, 'Why the Future Doesn't Need Us,' *Wired Magazine* (April 2000).
2 Jean Baudrillard, *The Vital Illusion* (New York: Columbia University Press, 2000), 20.
3 Ibid, 3.

Index

Digital Futures is a series of critical examination of technological development and the transformation of contemporary society by technology. The concerns of the series are framed by the broader traditions of literature, humanities, politics, and the arts. Focusing on the ethical, political, and cultural implications of emergent technologies, the series looks at the future of technology through the 'digital eye' of the writer, new media artist, political theorist, social thinker, cultural historian, and humanities scholar. The series invites contributions to understanding the political and cultural context of contemporary technology and encourages ongoing creative conversations on the destiny of the wired world in all of its utopian promise and real perils.

Series Editors:
Arthur Kroker and Marilouise Kroker

Editor Advisory Board:
Taiaiake Alfred, University of Victoria
Michael Dartnell, University of New Brunswick
Ronald Deibert, University of Toronto
Christopher Dewdney, York University
Sara Diamond, Banff Centre for the Arts
Sue Golding (Johnny de philo), University of Greenwich
Pierre Levy, University of Ottawa
Warren Magnusson, University of Victoria
Lev Manovich, University of California, San Diego
Marcos Novak, University of California, Los Angeles
John O'Neill, York University
Stephen Pfohl, Boston College
Avital Ronell, New York University
Brian Singer, York University
Sandy Stone, University of Texas, Austin
Andrew Wernick, Trent University

Books in the Series:
Arthur Kroker, *The Will to Technology and the Culture of Nihilism: Heidegger, Nietzsche and Marx*